AF433264

Harvest Haven

Creating Your Backyard Homestead Oasis

Chloe Nelson

© Copyright 2024 - All rights reserved.

The content contained within this book may not be reproduced, duplicated or transmitted without direct written permission from the author or the publisher.

Under no circumstances will any blame or legal responsibility be held against the publisher, or author, for any damages, reparation, or monetary loss due to the information contained within this book, either directly or indirectly.

Legal Notice:

This book is copyright protected. It is only for personal use. You cannot amend, distribute, sell, use, quote or paraphrase any part, or the content within this book, without the consent of the author or publisher.

Disclaimer Notice:

Please note the information contained within this document is for educational and entertainment purposes only. All effort has been executed to present accurate, up to date, reliable, complete information. No warranties of any kind are declared or implied. Readers acknowledge that the author is not engaging in the rendering of legal, financial, medical or professional advice. The content within this book has been derived from various sources. Please consult a licensed professional before attempting any techniques outlined in this book.

By reading this document, the reader agrees that under no circumstances is the author responsible for any losses, direct or indirect, that are incurred as a result of the use of information contained within this document, including, but not limited to, errors, omissions, or inaccuracies.

Table of Contents

INTRODUCTION

This intriguing book, "Harvest Haven: Creating Your Backyard Homestead Oasis," invites you to explore the fascinating world of farming. The desire to live sustainably, grow food, and connect with the environment is more vital than ever as urban landscapes expand. This e-book is your road map to turning your backyard into a thriving sanctuary where human endeavor and the natural world coexist peacefully.

A farm becomes a rewarding endeavor in the fast-paced world of today, offering not only healthful and fresh products but also a haven for mental and physical health. Homesteading can reduce stress, provide physical exercise, and improve diet quality. "Harvest Haven" is an invitation to set out on a path of self-sufficiency, environmental care, and community development rather than merely a how-to manual.

The chapters of this e-book have been carefully designed to give you the knowledge and abilities to establish a flourishing backyard retreat. Each part aims to provide practical and educational information, covering sustainable gardening practices, rearing happy livestock, and careful planning and plant selection.

You'll learn the delight of enjoying the fruits of your labor as we explore the fundamentals of homesteading, be it a plentiful crop of vegetables grown on your property, the fulfillment of raising contented animals, or the essential satisfaction of coexisting peacefully with the natural world.

"Harvest Haven" is designed to meet you where you are in your homesteading journey, no matter how experienced you are or how big or tiny your backyard is. Come learn how to create your own backyard homestead oasis that will not only provide for you and your loved ones but also help you develop a stronger bond with the land and the people in your neighborhood. Homesteading can also be a community effort where neighbors share resources, knowledge, and harvests. Prepare to transform your garden into a bountiful oasis where your efforts will yield concrete and profoundly satisfying results.

CHAPTER I

Assessing Your Space

Understanding Your Property

When you are working toward establishing a harmonious homestead oasis, one of the most important steps is to get a profound awareness of the land that you have available to you. There is more to your property than just a piece of land; it is a canvas just waiting to be painted on, a living ecosystem bursting with promise. To construct a prosperous homestead, this section aims to delve into the myriad of facets involved in grasping and optimizing your property.

Embarking on this journey, your first step is to conduct a comprehensive assessment of your available area. Recognizing your site's topography, soil composition, solar exposure, and microclimates is crucial. These parameters, which play pivotal roles, determine the types of plants that will thrive and the locations suitable for various activities. By understanding your land's topography in-depth, you can strategically plan your homestead layout, enhancing efficiency and productivity.

While evaluating your land, consider your homesteading goals. Are you aiming for a self-sufficient vegetable garden, a space for rearing animals, or a combination? Defining your goals is crucial as they guide your layout, infrastructure, and resource allocation decisions. Aligning your goals with your property's unique features increases the likelihood of your homesteading endeavor's success, whether you envision a thriving orchard, a cozy chicken coop, or a vibrant vegetable patch.

As soon as the more general objectives have been determined, it would help if you shifted your focus to the situation's specifics. Make sure you are familiar with the water drainage patterns on your land and locate any prospective spots prone to waterlogging or erosion. The implementation of efficient irrigation systems and the utilization of water-saving techniques become essential components of the property management approach that you utilize. Similarly, it is crucial to consider the direction of the wind since this can affect the microclimate and the growth of plants. You can strategically arrange windbreaks or select plant species that thrive in your particular conditions if you are aware of these nuances and consider them.

It is essential to pay particular care to the soil,

frequently referred to as the "heart" of a homestead. Perform a soil test to ascertain the composition of the soil, the amounts of nutrients, and the pH. You can now adjust the soil to create an environment that is ideal for the crops you have selected. When necessary, incorporate organic matter, compost, or other soil conditioners into the soil to cultivate a fruitful environment that encourages the rapid growth of healthy plants. To build a robust and productive garden, it is essential to have a solid understanding of the specific requirements of your soil.

Regarding homesteading, space is a material good that is quite valuable. Finding the regions of your property exposed to the most sunshine during the day is essential in becoming familiar with your land. This information is necessary to plan your garden beds' layout properly and determine the appropriate locations for animal cages. It is essential to ensure that plants receive the required sunlight to undergo photosynthesis and develop to their full potential. This can be accomplished by positioning the plants to align with the path of the sun.

Beyond your property's physical characteristics, consider the legal and regulatory aspects that could influence your homesteading plans. Research local zoning rules, obtain necessary permissions, and familiarize yourself with any restrictions. Understanding these parameters ensures compliance and helps you anticipate potential hurdles that could hinder your homesteading goals.

The process of gaining an awareness of your property extends beyond the confines of your land; it involves monitoring and engaging with the ecology in the surrounding area. It is essential to take notice of the native plants, the habitats of the fauna, and the overall ecology of the field. Because of this awareness, you are able to make decisions that are well-informed and contribute to the sustainable balance of the ecosystem rather than disrupting it. In addition, you should think about the effects that your homesteading activities have on the surrounding ecosystem and investigate different strategies to reduce the ecological imprint that you leave behind.

As you become more familiar with the specifics of your property, try to see it as a living, breathing organism that is constantly changing. With the changing seasons, your property's personality also shifts. Embrace the cyclical nature of homesteading by adjusting your goals and actions so that they are in harmony with the natural world's ebb and flow. You should be able to identify the subtle cues that mark the shifting of the seasons and use them to your advantage by organizing plants, harvests, and other activities with this information.

One of the most critical aspects of successful homesteading is solid awareness of the land surrounding you. Your land will be subjected to a comprehensive analysis considering its physical, ecological, and regulatory characteristics. With this information at your disposal, you can construct a homestead that not only complements the natural characteristics of your land but also satisfies your objectives and ambitions. As you continue on this road of comprehension and cultivation, it is essential to remember that your property is more than simply a piece of land; it is a canvas waiting for you to paint the colorful tapestry of your homesteading dreams.

Sunlight and Soil Analysis

The two essential components of agricultural productivity, sunlight, and soil, are closely linked to the prosperity of any homestead. This investigation explores how crucial it is to comprehend how these factors interact since they influence plant health, soil fertility, and a homestead's general sustainability.

Plants get most of their energy from sunlight, a life-giving substance pouring down from the sky. The food chain's basis is photosynthesis, which is how plants turn light into energy. Getting sunshine is one of many problems facing homesteaders; we also need to figure out how best to use it for the variety of plants we raise. Comprehending the dynamics of sunshine on your land while designing a strategic garden is essential.

Start by noting how your land is exposed to sunlight in different areas. Determine which regions are mostly shaded, receive some shade, or receive full sun. Because every plant needs additional sunlight, the arrangement of your garden beds can be guided by this knowledge. Some plants, like tomatoes, peppers, and squash, require full sun to grow well; on the other hand, many herbs and leafy greens may withstand some shade. Planting your trees in a way that best suits the patterns of sunlight on your land fosters development and productivity.

Think about how the amount of sunlight varies throughout the year. Observe the sun's path throughout the year to predict variations in the length and intensity of the light. With this information, planting periods may be adjusted for best results, and crops with different preferences for sunshine can be strategically placed. Furthermore, consider any possible obstacles, such as trees or buildings, that could shade your garden.

Reflective surfaces can be deliberately positioned or pruned to reduce shade and increase the accessible sunshine.

The intricate universe of dirt beneath our feet balances the sun's dance. A dynamic ecosystem brimming with microbes, minerals, and organic matter, soil is not a static object. A thorough soil analysis is vital for determining your property's unique requirements and adjusting your homesteading methods accordingly.

Start your soil analysis by learning about the three main constituents: clay, silt, and sand. Soil texture is determined by the makeup of these particles, which also affects drainage, aeration, and water retention. Clayey soils hold water but can become compacted, whereas sandy soils drain fast but do not have enough nutrients. In general, loamy soils are balanced and perfect for plant growth. Understanding the texture of your soil helps you select crops that will thrive there and add nutrients to improve the structure of the soil.

Take a soil test to learn more about the nutrients in your soil. This research highlights essential components, including micronutrients, potassium, phosphorus, and nitrogen. Equipped with this knowledge, you may customize your fertilizer treatments to target particular inadequacies and preserve a nutrient-rich atmosphere for your plants. Composted or well-rotted manure are examples of organic amendments that enhance soil structure and add nutrients, creating a healthy and productive base for your farm.

The pH of the soil, a gauge of its acidity or alkalinity, significantly impacts the availability of nutrients. With a pH range of 6.0 to 7.0, most plants grow best in neutral to slightly acidic soils. You can determine whether your soil is in this ideal range by testing its pH. Too high or too low of a pH can prevent plants from absorbing nutrients. While elemental sulfur is typically administered to lower pH in alkaline soils, lime is frequently used to boost pH in acidic soils. Keeping the pH at the proper level guarantees that your plants get the nutrients they require for healthy growth and development.

Please take into account the biological aspect of your soil in addition to its chemical features. Earthworms, fungi, and bacteria are a few microorganisms that comprise a healthy soil ecosystem. These organisms are essential for cycling nutrients, enhancing soil structure, and controlling pests. A healthy soil microbiome is supported by cover crops, mulching, and little soil disturbance, which increase your homestead's overall resilience.

Examining the idea of microclimates makes the relationship between sunshine and soil clear. Microclimates are small-scale climatic differences brought about by the makeup of the soil, vegetation, and topography. Comprehending the microclimates on your land enables you to plant crops according to their particular needs. A sunny, well-drained slope may be perfect for heat-loving plants, while crops sensitive to high temperatures may do better in a protected, more relaxed spot.

Take a comprehensive approach that considers the interactions between sunlight and soil analyses as you include them in your homesteading methods. For example, crop rotation becomes a practical approach when the particular qualities of your land drive it. Plants with varying nutrient requirements are cycled through specific locations to prevent nitrogen depletion and lower the danger of soil-borne illnesses.

In addition, adopt sustainable methods that improve the ecosystem's equilibrium on your farm. By integrating trees and shrubs into the agricultural environment, agroforestry can produce windbreaks, shade, and extra organic matter from falling leaves. These components facilitate a robust and self-sustaining homestead environment.

In summary, the skillful comprehension and management of sunlight and soil constitute the foundation of homesteading expertise. These components, commonplace in their abundance, are the keys to realizing your land's full potential. Let the sun's dance direct your garden design as you set out to create your little haven on your property, and use the earth under your feet as a blank canvas to create the colorful tapestry of your produce. The secret to a plentiful and sustainable homesteading experience is found in this complex interaction between light and earth.

Zoning and Legal Considerations

Embarking on the journey of creating a homestead is rewarding, yet it demands more than just a love for nature and a knack for gardening. An often overlooked but crucial aspect in cultivating a self-reliant haven is comprehending the zoning restrictions and legal considerations that govern your land. This section sheds light on the importance of compliance, responsible land use, and the potential hurdles aspiring homesteaders might face. It does so by delving into the intricacies of zoning and legal considerations.

Zoning restrictions, set by local governments, dictate the allowable uses of land within a specific jurisdiction. Land areas are categorized into residential, commercial, agricultural, or industrial use based on these regulations. As a homesteader, you must acquaint yourself with the zoning rules applicable to your property. This understanding ensures your compliance with the law and empowers you to make informed decisions about the layout, constructions, and activities on your land.

The first step in navigating zoning restrictions is to secure a copy of the zoning map and law relevant to your area. Typically, zoning maps illustrate various zones and the designations and legal uses of property corresponding to each category. Residential zones often allow small-scale gardening and agriculture, but the specifics of this allowance vary from place to place.

Agricultural zones offer more leeway for farming activities, yet restrictions may apply. To align your homesteading goals with the legal uses outlined in the zoning regulations, it's essential to grasp these nuances.

The acquisition of a particular use permit or a variance may be required in some circumstances to participate in activities that are not expressly permitted by the zoning that is currently in place. Activities that depart from the conventional land-use restrictions but are deemed acceptable under specific conditions are often eligible for special use licenses. These permits are typically provided for activities. In contrast, variances offer relief from particular zoning regulations imposed on a property because of the one-of-a-kind conditions associated with the property. On the other hand, the license application process can be complicated since it involves public hearings and approvals. Because of this, it is vital to confer with the local planning authorities.

Criteria for the location and building of structures are frequently included in zoning regulations, which control how land is used and fits these criteria. Common zoning issues include setback regulations, which specify the minimum distance that structures must be from the land's boundaries or from adjacent structures. To plan the layout of your homestead, it is essential to have a thorough understanding of these setbacks. This will ensure that your buildings and gardens are by the prescribed distance. If setback standards are not adhered to, an individual may be subject to penalties or, in the most severe circumstances, the necessity to alter or relocate structures.

In addition to zoning rules, environmental regulations, building codes, and health and safety requirements were also considered by the legal system. Homesteading operations must operate inside a framework shaped together by these elements. Regulations about the environment may address concerns such as the preservation of wetland areas, the utilization of water, or the protection of endangered animals. The purpose of establishing building codes is to provide structural integrity, fire safety, and accessibility standards for the construction industry. Compliance with these regulations is a legal requirement and protects the health and safety of the homesteader and the community.

Homesteaders must also take into account the fact that water rights and usage are essential legal factors. It is crucial to have a solid understanding of the regulations that control water rights in your respective location, mainly if your homesteading activities entail irrigation, animals, or crops that require a significant amount of water. Water to the property in certain regions, while in others, they may be subject to separate licenses. If you know the regulations governing water, you can make responsible use of this valuable resource while adhering to established legal parameters.

All applicable health and safety regulations should do the incorporation of animals into your property. To prevent the transmission of diseases, ensure that animals are treated humanely, and effectively manage waste, livestock management may be subject to special rules. Make sure you are familiar with the local regulations regarding animal management, and make sure you secure any permissions that are required for keeping animals. This proactive approach not only helps to avoid legal concerns but also helps to create a homesteading practice that is responsible and sustainable within the community.

To successfully navigate the legal landscape, you must maintain active engagement with the community and the authorities in your area. It is possible to reduce the likelihood of potential disputes by cultivating positive connections with neighbors and remaining knowledgeable about the community's expectations. Regarding grasping the legal framework and connecting with individuals with similar interests, local agricultural extension offices, planning departments, and community organizations can be beneficial tools.

You need to keep your awareness current with the ever-changing legal situation. Maintain awareness of any modifications in zoning restrictions, environmental laws, or other possible legal concerns that may impact your homestead. Participate in meetings of the local government, participate in community forums, and have an open line of communication with the appropriate authorities and authorities. By taking this preventative strategy, you will be able to quickly adjust to any changes in the law and ensure that your homesteading activities will continue to be lawful.

CHAPTER II

Planning Your Oasis

Designing Your Homestead Layout

A homestead's layout is like a canvas on which aspirations of sustainability and self-sufficiency are painted. It's a detailed dance of planning and imagining that considers the homesteader's objectives, the land's unique qualities, and the mutually beneficial relationships between different components. This study delves into the art and science of planning a homestead layout, which entails thoughtful space planning, effective resource management, and the development of a peaceful atmosphere that promotes well-being and productivity.

Assessing the available area in detail is one of the first steps in creating a layout for a homestead. Recognize the land's geography, including any hills, valleys, or other distinctive features that could affect solar exposure or water drainage. This information serves as the cornerstone around which the layout is constructed, enabling the thoughtful arrangement of various components like garden beds, animal cages, and buildings.

Take your homestead's objectives into consideration as you start this planning process. Are you picturing a vegetable garden that provides most of the vegetables for your family? Will you include livestock on your farm to promote a mutually beneficial link between plants and animals? Setting clear goals is crucial because they influence how things are arranged and how much room is allotted to different activities.

Based on your aims, create functional zones on your property. Making separate zones for living areas, animals, and gardening is a popular strategy. Allocate different crop zones depending on the amount of sunlight and water available. If your homestead idea includes animals, design their living quarters with things like waste management, grazing places, and shelter in mind. Coherent integration of these zones guarantees efficient workflow and reduces needless movement throughout the homestead.

Effective use of available space is a symptom of a well-planned homestead layout. For instance, vertical gardening uses vertical structures like trellises, hanging baskets, or stacked containers to optimize growing space. This strategy is effective, particularly in smaller homesteads where every square foot matters. Furthermore, companion planting maximizes space and improves the general health of the garden by growing several crops together depending on their mutually beneficial associations.

Another critical component of homestead design is the positioning of structures strategically. Place buildings like sheds, greenhouses, or barns to use as much natural light as possible and make entry simple. When placing these components, consider potential sources of shade and the orientations of the prevailing winds. A well-planned layout improves the homestead's aesthetics and efficiency while fostering an aesthetically pleasant and helpful atmosphere.

One of the most critical factors in designing a homestead is water management. Make plans for effective irrigation systems to meet your crops' needs while saving water. Soil erosion can be avoided, and water runoff can be managed by implementing contour planting, swales, or rainwater collection systems. Water is distributed evenly throughout your property thanks to the thoughtful placement of irrigation systems and water sources, encouraging strong plant growth and reducing waste.

Create walkways as part of your homestead design to make moving around and accessing various zones easier. Consider using porous materials to encourage water absorption and reduce runoff for paths. In addition to being functional, pathways also enhance the homestead's overall beauty. Visitors can be guided around your homestead's numerous components via well-designed routes, providing an opportunity for exploration and a deeper connection to the land.

A key component of homestead design is companion planting, a method based on the symbiotic connections between various plants. Think about how some plants complement one another; for example, one species may encourage the growth of another or ward against pests. For instance, growing fragrant herbs like basil next to tomatoes helps ward off pests that frequently damage them. This well-planned layout makes the most of available space and encourages a natural equilibrium in the garden, lessening the need for chemical treatments.

A key component of many homesteads is the integration of livestock, and creating areas for animals necessitates considering their needs and habits. Provide animals with comfortable and helpful living environments by adding amenities such as safe enclosures, eating spots, and shelter. If appropriate, schedule grazing in a rotation to provide pastures time to recover and rejuvenate. Another essential factor to take into account is waste management. Well-thought-out plans include methods for composting and using animal manure as a resource for the garden.

When designing your homestead, take permaculture ideas into account. Based on regenerative and sustainable methods, permaculture promotes the development of systems that resemble natural ecosystems. To build your homestead layout according to permaculture principles, you must consider and comprehend the natural cycles and patterns. The resilience and sustainability of the homestead are enhanced by features like polyculture, the coexistence of various crops under one roof, varied plantings, and the inclusion of perennial crops.

Consider aesthetics when creating your homestead layout and aim to develop areas that uplift and revitalize. Incorporate decorative flowers, plants, and landscape elements that enhance the homestead's aesthetic appeal while drawing pollinators and beneficial insects. Create outdoor living areas where you may unwind, enjoy the natural world, and acknowledge the results of your hard work. A well-planned homestead is a haven that uplifts the soul in addition to being a practical area.

The design of a homestead must also consider legal factors, especially regarding zoning compliance. Verify that the design complies with local zoning regulations and get the required permissions for any constructions or activities requiring regulatory approval. Knowing the law from the beginning is best to avoid problems or disruptions to your homesteading experience.

In summary, creating a homestead layout is a dynamic and imaginative process that combines practicality and creativity. It calls for careful planning, a deep knowledge of the soil, and a dedication to regenerative and sustainable methods. In addition to assisting with the practical concerns of raising food and caring for animals, a well-planned homestead layout promotes a strong bond with the land and a sense of balance with the natural world. When planning the layout of your homestead, listen to the earth and let your vision come to life in sync with the cycles of the natural world.

Selecting Crops and Livestock

Any homestead's core is its meticulous selection of animals and crops, a process that calls for planning, forethought, and an awareness of the unique requirements and dynamics of the land. Homesteaders' decisions when choosing their animals and crops become crucial as they set out to build a successful and sustainable agricultural paradise. This section delves into the complex relationship between the homestead's flora and fauna, illuminating factors like soil health, climate compatibility, and the interaction of various components that lead to a self-sufficient and flourishing homestead.

Climate compatibility is the primary factor when choosing crops for a farm. Because different plants grow better in different climates, crops should be selected according to the region's sunlight patterns, precipitation, and temperature. Knowing your area's USDA hardiness zone is a great way to determine which crops will most likely thrive. While cold-hardy vegetables like kale, carrots, and broccoli flourish in colder regions, heat-loving crops like tomatoes, peppers, and eggplants are best suited to warmer temperatures. Homesteaders can prepare for succession plantings and lengthen the growing season by taking advantage of seasonal fluctuations and frost dates, further improving the choosing process.

Crop cultivation is predicated on the health of the soil. Performing a comprehensive soil study reveals the soil's nutritional makeup, pH levels, and texture, offering information that affects crop choices. Certain crops require particular soil types; for example, leafy greens like lettuce and spinach grow best in loamy, nutrient-rich soil, while root vegetables like potatoes and carrots prefer loose, well-drained soil. Crop rotation combined with cover crops improves soil fertility and structure, reducing the chance of nutrient depletion and fostering a robust and healthy homestead ecosystem.

One important consideration when choosing crops for a farm is diversity. Adopt a polyculture strategy and grow a range of mutually beneficial crops. Companion planting is a long-standing technique that promotes biodiversity, pest control, soil fertility, and general garden health. It is based on the advantageous interactions between various plant species. For instance, putting fragrant herbs like basil next to tomatoes can help create a mutually beneficial environment by keeping pests away that typically damage tomatoes.

When choosing crops, take your household's nutritional requirements into account. Grow various fruits, vegetables, and herbs to ensure a wholesome and balanced diet. Including crops that store well, like winter squash, potatoes, and onions, guarantee a consistent food supply all year. By preserving traditional plant genetics and promoting biodiversity, experimenting with heirloom and open-pollinated varieties helps people feel more connected to the past of agriculture.

Crop choices are significantly influenced by the dimensions and design of your farm. Vertical gardening becomes a valuable method to increase growing areas in cramped areas. Use raised beds, trellises, and hanging pots to maximize the restricted space. Another choice for people with little space is container gardening, which offers mobility and placement flexibility. Make sure your crop choices are appropriate for the available space to maximize production and resource efficiency.

When choosing livestock, homesteaders should consider each animal's unique requirements, spatial constraints, and planned uses. Adding livestock to a farm creates a dynamic relationship wherein the animals help with pest control, soil fertility, and social relationships. For example, hens produce eggs, naturally manage insects, and leave behind nutrient-rich dung that benefits gardens.

The type of animals chosen should fit the homestead's objectives and size. Micro-livestock, such as backyard hens or rabbits, might be more appropriate in tiny areas and provide a pleasant but manageable experience. Larger animals, such as goats, sheep, or even pigs, may be allowed on larger farms; each has advantages and disadvantages of its own. Pay attention to local zoning laws that can affect the kinds and quantity of animals permitted on your land.

When choosing cattle, take your region's climate and geography into account. Selecting animals that flourish in your surroundings enhances their well-being and the success of your homestead. Certain breeds are more suited to particular climates than others. While heat- tolerant breeds are better adapted to warmer areas, cold-hardy breeds may be more adaptable in northern settings. Evaluate the availability of grazing spaces and forage to satisfy the animals' nutritional requirements.

Include cattle in the household ecology that fulfills several functions. Goats, for instance, are excellent at removing bush and yielding milk and meat, which makes them useful for land management. By feeding on insects, snails, and weeds, ducks help manage pests while also producing eggs. Because of their innate tendency to root, pigs can be used to clear weeds from garden beds.

It is essential to comprehend the social dynamics and behavior of the selected cattle. While some animals love living in groups, others would instead be left alone. Consider the need for shelter and space and the possibility of interspecies conflict or coexistence. A well-thought-out layout for animal housing and fencing assures the livestock's safety and welfare by limiting escape and lowering stress.

Use rotational grazing techniques to minimize overgrazing and maximize pasture health. This entails regularly relocating cattle to new locations so pastures can heal and regress. As the animals graze on new, nutrient-rich fodder, the cycle improves the land and the animals' health. Using a controlled grazing system and separating pasture lands into paddocks enhances soil health and sustainability in more significant homesteads.

Think about the resources and long-term commitment needed to raise animals. Animals require regular feeding, care, and attention to their health. Create a thorough strategy that addresses seasonal concerns, nutrition, and veterinarian care. Have backup plans in case of emergencies or other unforeseen problems, such as protecting yourself from predators. Building a strong bond with your livestock increases mutual trust and improves the health of the homesteader and the animals.

In summary, choosing animals and crops for a farm requires a careful balancing act between the pragmatic and the visionary. Every choice you make affects the homestead's overall resilience and success, from promoting biodiversity and sustainable practices to comprehending soil dynamics and climate. The procedure is dynamic and changes as the seasons do, catering to the requirements of the land and the homesteader. Let the crops and livestock you choose as you begin your cultivation and stewardship reflect your values, goals, and dedication to building a peaceful and sustainable farm.

Creating a Sustainable Watering System

Water is essential to any homestead because it supports cattle, feeds crops, and creates a vibrant ecology that turns a plot of land into an oasis. Creating a thoughtfully planned and effective irrigation system becomes essential for homesteaders who want to foster sustainability in all activities. This section examines the many aspects of designing a sustainable irrigation system, from collecting and using rainfall to implementing resource-saving irrigation techniques that promote the homestead's long-term health.

The foundation of a water-efficient farmhouse is the creative and sustained practice of harvesting rainwater. Gathering and storing precipitation runoff from roofs or other surfaces for future use is known as rainwater harvesting. Installing a catchment system, which usually uses gutters and downspouts to direct rainwater into storage containers, is the first step in the process. Rainwater collection is stored in barrels, cisterns, or other storage containers to provide a dispersed and environmentally beneficial water source.

Rainwater gathering has advantages beyond water resource conservation. Homesteaders can lessen their need for outside water supplies, alleviate the effects of stormwater runoff, and help control soil moisture by collecting rainwater. Rainwater is the best option for irrigation since it is gentle by nature and doesn't include any of the pollutants frequently present in municipal water systems. The technique encourages self- sufficiency and lessens the homestead's ecological footprint, all in line with sustainability concepts.

Water demand, distribution strategies, and storage capacity must all be carefully considered when integrating rainwater collection into a household. To find the right size storage containers for your needs, evaluate the amount of rain that falls there on average each year. Determine how much water your cattle and crops need, accounting for seasonal fluctuations, and then build the system to satisfy their demands. Low-energy pumps or gravity-fed distribution systems guarantee eco-friendly and efficient water flow.

A focused and water-efficient way to send moisture straight to a plant's roots is through drip irrigation. This technique is a fantastic option for sustainable homesteading since it decreases water waste by lowering runoff and evaporation. By using tubes or hoses with emitters that release water gradually and steadily, drip irrigation ensures that plants receive a constant water supply without having their leaves overly damp. The method is flexible and adaptable to diverse homestead demands because it can be tailored to suit different crop varieties and layouts.

As part of your sustainable watering system, consider using soaker hoses or porous pipes in addition to drip irrigation. These systems allow for the gradual and even irrigation of plants by releasing water along their length. For garden beds, soaker hoses work exceptionally well since they minimize water contact with leaves while hydrating plants' root zones. Adequate and targeted hydration, soaker hoses, and drip irrigation promote water conservation and plant health.

Mulching techniques are another way to improve a homestead's water efficiency. Mulch forms a protective layer on the soil's surface and can be made of wood chips, straw, or organic compost. It slows the rate of evaporation, inhibits the growth of weeds, and keeps the soil from drying up too quickly. Mulching is an easy-to-use but efficient method that helps the homestead become more sustainable overall, improves soil health, and conserves water.

Soil moisture monitoring becomes essential for precision irrigation when constructing a sustainable watering system. To determine the soil's moisture content, use manual techniques such as the finger test or moisture sensors. Making decisions about when and how much water to apply is made more accessible by this information, which also helps to ensure that plants get the right amount of moisture. Homesteaders maximize water use and reduce waste by matching irrigation techniques to the unique requirements of the land and crops.

Consider adding rain sensors to your irrigation system to improve water efficiency. These sensors detect rainfall and momentarily halt the irrigation schedule to avoid watering unneeded plants during or right after a rain event. Smart irrigation controllers further automate watering by modifying schedules based on current conditions. These controllers are outfitted with weather-based or soil moisture-based algorithms. By minimizing dependency on set schedules and optimizing irrigation time, these devices help conserve water.

The kind of crops, the homestead's design, and water availability all influence the choice of suitable irrigation techniques. Sprinkler systems typically serve larger areas than soaker hoses and drip watering. To reduce water wastage, it is crucial to select sprinkler heads that use less water and to arrange them optimally. The best sprinklers are those that rotate or oscillate to disperse water evenly. The use of automated controllers or timers ensures accurate control over irrigation schedules.

A sustainable farm must include livestock watering, and an efficient water supply design promotes animal welfare and efficiency. Installing mechanical waterers or gravity-fed water systems minimizes the need for physical labor and guarantees a steady supply of clean water for cattle. Utilizing rainwater to hydrate cattle is another way to use water collection systems for animal watering sites to improve sustainability.

P.A. Yeomans popularized the idea of keyline design, an all-encompassing method of managing water resources that considers the land's natural contours. Finding the keyline, a contour line indicating the best route for water to flow across the landscape, is necessary for this method. Homesteaders can capture and direct precipitation by creating swales along the keyline, inhibiting runoff and encouraging soil infiltration. Retaining soil moisture, using water sustainably, and building overall landscape resilience are all facilitated by boundary design.

An advanced tactic for homesteaders with more significant acreage and water features is wetlands restoration. Homesteaders boost biodiversity, water quality, and natural water storage systems by farming and restoring wetland areas. Wetlands function as sponges, taking in and purifying water, preventing flooding, and providing habitat for various species. This water management method is holistic and regenerative, even if it involves careful planning and ecological considerations.

Utilizing graywater systems allows for the sustainable repurposing of home water for irrigation. Graywater can be cleaned up and used again in the garden. This includes water from sinks, showers, and washing machines. Putting in place a graywater

The system complies with recycling and resource efficiency guidelines while lowering the strain on freshwater resources. However, when establishing graywater systems, it's imperative to follow municipal laws and ordinances.

Homesteaders must first consider ecological harmony, water efficiency, and conservation when designing a sustainable irrigation system. Homesteaders contribute to the larger objective of building a resilient and sustainable environment and ensuring the health of their crops and cattle by collecting rainwater, using effective irrigation techniques, and implementing regenerative activities. A water-wise farmhouse embodies the values of responsible stewardship and peaceful coexistence with the natural world. An ongoing process of observation, adaptation, and a close relationship with the land reaches it.

CHAPTER III

Essential Tools and Equipment

Tools for Every Homesteader

With its allure of sustainable living and self-sufficiency, homesteading is an undertaking that requires commitment and a good toolkit. The proper tools are essential for every part of this varied lifestyle, from tilling the soil and caring for cattle to constructing buildings and preserving the homestead ecology. This section examines the necessary instruments that any homesteader should take into account. These tools range from simple hand tools to sophisticated equipment, and each is vital to the development of a sturdy and prosperous homestead.

A homesteader's constant companions are fundamental hand tools, which are the foundation of any homestead. To prepare the soil, sow seeds, and maintain garden beds, you'll need a trustworthy set of gardening equipment, such as a substantial shovel, a flexible hose, and long-lasting rakes. Hand pruners and shears are essential tools for pruning plants, trimming shrubs, and gathering fruit. These instruments, frequently the most basic in construction, serve as the foundation for the homesteader's everyday activities and promote a close relationship between the hands and the land.

A well-made, sharp knife is a homesteader's best friend, helpful for anything from harvesting vegetables in the garden to tending to livestock. A good knife is essential for every homesteader's toolbox, whether it's for processing newly gathered vegetables, trimming unwanted growth in the garden, or even helping with meal preparation. A multipurpose pocket knife is also helpful for various activities around the homestead, providing efficiency and ease in day-to-day duties.

As they progress beyond the fundamentals, homesteaders frequently discover that they need power equipment to do increasingly tricky jobs. For instance, a chainsaw is valuable for handling trees, clearing land, and preparing firewood. When tackling various cutting tasks, efficiency and safety are ensured by choosing a chainsaw with the right power and bar length. A chainsaw that runs on gas or electricity is adaptable, providing power and mobility for various uses.

A cordless drill is essential for any homesteader due to its versatility. Whether for construction, fencing assembly, crafting, or maintenance, a cordless drill makes jobs that would typically need much labor much more accessible. A single cordless drill can do multiple tasks, such as driving screws, drilling holes, and providing power for additional accessories by using interchangeable bits.

Whether erecting a greenhouse for year-round gardening, a barn for cattle, or a coop for hens, homesteaders are frequently involved in construction projects. A sturdy circular saw is essential for accurately and effectively cutting various materials, such as plywood and lumber. When equipped with the appropriate blade, a circular saw can be used for multiple jobs in general construction, woodworking, and framing.

A sturdy and robust tractor can be a game-changer for homesteaders who manage more significant agricultural areas. Tractors, the workhorses of the homestead, were used for various chores like tilling soil, plows, and heavy lifting. Adding attachments to your tractor, like plows, harrows, and tillers, becomes even more versatile and is a necessary investment for effectively managing larger homesteads.

A unique collection of instruments designed with the welfare of livestock in mind is required for their care. Hoof trimmers are essential for preserving the well- being of sheep, goats, and other animals with cloven hooves. Hoof trimming done correctly keeps animals from becoming lame and uncomfortable, which improves their general welfare. Some cattle breeds can also need dehorning equipment to avoid harm and guarantee safe handling.

Fencing is essential to control cattle and define distinct areas on the homestead. Homesteaders can construct garden borders or safe animal enclosures using a post-hole digger to make installing strong fence posts easier. This instrument is necessary to build sturdy and dependable fences, whether used for significant projects with a power drill or smaller ones with a manual post-hole digger.

Homesteaders benefit immensely from having an accurate and weather-resistant thermometer because of the constant risk of unpredictable weather and changing seasons. Keeping an eye on temperature swings is critical to safeguard cattle, crops, and even beehives' comfort. Homesteaders can make well-informed judgments based on the present weather conditions by using real-time data from a digital thermometer equipped with remote sensors.

A pressure canner is essential for homesteaders learning the craft of storing and preserving their produce. This specialist canning equipment may safely keep vegetables, meats, and soups. A pressure canner, which works at a temperature greater than conventional water bath canners, guarantees the eradication of hazardous bacteria, prolonging the shelf life of preserved foods and enhancing domestic food security.

Many homesteaders have taken up the age-old hobby of beekeeping, which calls for specialized equipment to maintain beehives and gather honey. A hive tool is necessary for removing frames, scraping off extra wax or propolis, and pulling apart hive components. During hive inspections and honey extraction, the beekeeper's veil and gloves offer protection, guaranteeing a secure and pleasurable beekeeping experience.

A good pruning saw is handy when homesteaders start growing fruit trees and orchards. Pruning is necessary to shape trees, eliminate unhealthy or dead branches, and encourage strong, healthy growth. A pruning saw's design makes cutting well in confined spaces possible, making it a valuable instrument for maintaining orchards. Adding hand pruners and loppers to a pruning saw completes the homesteader's tree-care toolset.

An electric fence energizer that runs on solar power is helpful for homesteaders who prioritize renewable energy sources for controlling animals and maintaining boundaries. This gadget gives electric fences a constant power source by using solar panels to charge a battery. Eco-friendly and economical, solar-powered electric fence energizers provide a reliable way to secure property.

A soil pH meter becomes a vital tool for homesteaders looking to maximize crop health. It is possible to regulate the pH of the soil precisely to produce a growing environment for plants by keeping an eye on its levels. The proper soil pH must be maintained, whether growing decorative plants, fruits, or vegetables, for the plants to be nutrient-available and healthy overall.

A dependable water pump is a valuable tool for homesteaders to manage their water resources. A water pump makes efficient water distribution possible for animal and crop irrigation or water extraction from ponds or wells. A homestead's water supply can be sustained by using a variety of water pumps, including submersible and jet pumps, designed to meet specific requirements and water sources.

Building a successful homestead is a continuous process that calls for commitment, flexibility, and the appropriate equipment. Every instrument, no matter how big or small, has a purpose in the many chores that homesteaders perform daily. When building a homestead, homesteaders must remember that having the proper tools is just as important as knowing how to use them and how they fit into the larger picture of what makes a resilient and sustainable homestead.

Machinery and Equipment for Efficiency

The wise use of machinery and equipment becomes essential to productivity, sustainability, and efficiency in homesteading, where living is in harmony with the soil. Initially limited to manual work and hand tools, the modest farmhouse now includes a variety of machinery intended to improve productivity, optimize resources, and expedite operations. This section explores the different roles that machinery and equipment play in various elements of construction, livestock management, cultivation, and general homestead efficiency. It digs into the machinery and equipment that have become essential to the modern homestead.

On the homestead, the tractor is a widely recognized representation of strength and efficiency. From its agricultural beginnings, this adaptable machine has become an essential tool for many homesteading duties. The tractor is the epitome of versatility—it can be used for everything from tilling soil and field preparation to towing oversized cargo and managing several attachments. When a front-end loader is added, the tractor becomes a powerful lifting and hauling machine that may help with construction operations and transportation of grain and commodities. The tractor's usefulness also extends to land management, where tools like cultivators and harrows help homesteaders prepare the soil for planting, making farming more efficient and productive.

During harvest season, a combine harvester is a game-changer for homesteaders who embrace more extensive tracts of land or intensive farming. This mechanical marvel effectively reduces the labor-intensiveness of manual harvesting by reaping, threshing, and winnowing crops like grains. The capacity of the farmhouse to process vast amounts of crops with little assistance from humans is increased by the combine harvester, which is an advantageous tool for farmers growing large fields of wheat, corn, or other grains.

The hay baler is essential for gathering and storing animal feed in livestock management. Hay is an integral part of winter feed for many homesteads, and the baler streamlines the hay-gathering, hay-compressing, and hay-packaging process into easily handled bales. This speeds up the process of creating hay and helps with adequate livestock feeding and storage. Using a baler ensures that animals have access to wholesome fodder all year round by the concepts of resource optimization.

Homesteaders frequently use the manure spreader to solve the problem of managing manure. This equipment improves soil fertility and reduces the need for artificial fertilizers by making it easier to spread nutrient-rich manure over fields. Effective manure spreading encourages good land stewardship and increases the sustainability of the homestead by reusing organic waste. The application of a manure spreader is a prime example of how machinery may be incorporated into ecologically conscious and efficient holistic farming techniques.

Homesteaders are adopting solar-powered water pumps at an increasing rate as they look for renewable energy sources. These solar-powered devices deliver a reliable and environmentally responsible way to distribute water for general use, livestock watering, and irrigation. Solar-powered water pumps provide homesteaders with an off-grid option that complies with the values of environmental responsibility and self-sufficiency by lowering dependency on traditional electricity sources.

Homesteaders maximize their ecological footprint while improving their water management techniques by using solar energy to operate water pumps.

The chainsaw is a mainstay in the homesteader's arsenal, which epitomizes effectiveness in managing wood resources. Whether for building projects, clearing land, or making firewood, the chainsaw speeds up operations that would otherwise need a lot of physical labor. Homesteaders now have even more options thanks to the introduction of quieter, greener chainsaws that run on electricity or batteries. In addition to meeting the homestead's energy demands, effective wood resource processing advances sustainable land management techniques.

Fruit pickers are a valuable addition to the toolset for homesteaders learning the art of orcharding and fruit cultivation. This easy-to-use yet efficient equipment reduces the possibility of damaging fragile branches and fruits when collecting fruits from tall trees. Homesteaders can efficiently and productively gather fruits like apples, pears, and plums using a fruit picker's extension pole and soft basket design. This instrument exemplifies how machinery, even in its most basic forms, may improve productivity and streamline the harvesting process.

A portable sawmill changes how homesteaders approach producing lumber in the construction industry. This small, lightweight device makes it possible to mill logs into lumber on-site, offering building projects an economical and environmentally friendly option. Utilizing wood from one's property encourages a more self-sufficient approach to building and lessens dependency on lumber supplied commercially. Homesteaders may now efficiently and precisely harvest, process, and use wood resources thanks to the portable sawmill.

Many people believe greenhouses are necessary to lengthen the growing season, and they have experienced advancements in both design and technology. Hoop houses and high-tunnel greenhouses are becoming more and more common on modern homesteads since they are affordable substitutes for more conventional buildings. These enclosed growing areas maximize natural light and heat while offering plants a regulated environment thanks to using materials like PVC pipes and plastic sheeting. Hoop houses are influential because they are easy to use, reasonably priced, and can help with year-round farming.

A dairy enterprise can be transformed entirely by an automated milking system for homesteaders prioritizing animal husbandry. These mechanical milking systems, built for both efficiency and animal welfare, enable timely and consistent milking without the need for human involvement. In addition to maximizing labor resources, automatic milking systems also improve the general health of dairy cows. Using technology in animal husbandry reflects the homesteader's dedication to efficiency and adherence to moral standards in caring for their cattle.

With the introduction of innovative farming solutions, the wide range of homesteading tools and machinery is no longer limited to the field or barn. Instead, it is now part of the technological landscape. Homesteaders may get real-time data and automation features through soil sensors, weather monitoring systems, and smart irrigation controls. Water management can be done precisely with these technologies, enabling the best irrigation practices depending on soil characteristics and weather predictions. Innovative farming technologies improve the homesteader's capacity for data-driven decision-making, supporting sustainable practices and resource efficiency.

In summary, the incorporation of machinery and equipment within homesteading's overall structure signifies a The subtle balancing act between innovation and tradition. Although hand tools are still a reliable partner for everyday tasks, introducing equipment has increased the homesteader's ability to be sustainable, productive, and efficient. Every piece of equipment, from solar-powered water pumps and automated milking systems to tractors and combine harvesters, contributes to the development of the modern farmhouse. Careful selection, prudent application, and a steadfast dedication to sustainability principles are essential in making machinery a cooperative partner in building a sturdy and prosperous homestead.

Budget-Friendly Options

With its allure of sustainable living and self-sufficiency, homesteading frequently evokes visions of picturesque scenery and abundant crops. But starting and keeping up a homestead might be relatively inexpensive. In their quest for a more balanced and economical existence, Homesteaders frequently look for affordable solutions that support their goals of living a more straightforward and independent lifestyle. This section examines several options homesteaders might choose wisely and economically without sacrificing the fundamentals of resilience and sustainability.

The soil that a homestead is built upon is its foundation. Spend okay to establish rich, healthy soil is unnecessary. Composting is one of the most affordable and environmentally responsible ways to improve soil fertility. Reusing organic matter, yard waste, and food scraps to create a compost pile enables homesteaders to replenish soil nutrients. Additionally, composting is made more affordable by obtaining free or inexpensive organic materials from nearby farms, such as manure, grass clippings, and leaves. Homesteaders practice sustainability and thrift simultaneously as they lessen their dependency on store-bought fertilizers through composting, which turns trash into a beneficial soil additive.

Using open-pollinated and heritage seeds is another cost-effective tactic for homesteaders who want to cultivate various crops. Heirloom seeds provide the benefit of seed saving, in contrast to hybrid seeds, which would need to be purchased again every season. By gathering seeds from mature plants, homesteaders can guarantee a perpetual growth cycle without ongoing expenses. The homesteading community can increase their seed repertoire at little cost by engaging in local seed exchanges or exchanging seeds. Homesteaders reduce the monetary outlay for yearly seed purchases while preserving traditional plant genetics and enhancing biodiversity by tending to a heritage garden.

Many homesteads are characterized by raising poultry, which provides eggs, meat, and organic pest management. Selecting breeds that produce meat and eggs provides homesteaders an affordable way to raise chickens. Certain breeds, including Sussex, Plymouth Rocks, and Rhode Island Reds, are noted for their hardiness and adaptability, which minimizes the need for specialist breeds and the expenses involved in keeping distinct flocks. Moreover, buying day-old chicks from nearby hatcheries or homesteaders is frequently less expensive than purchasing mature birds. Homesteaders can create a financially responsible and sustainable chicken farm by prioritizing dual-purpose breeds and economical sourcing.

When it comes to infrastructure, rainwater collecting shows up as an affordable way to guarantee a steady supply of water for the household. For a variety of purposes, such as watering plants or raising livestock, homesteaders can collect and store rainwater by installing straightforward rain barrels beneath gutter downspouts. Over time, this cost-effective method of managing water can result in significant savings on water bills and less reliance on municipal water supplies. To further reduce early investment costs, DIY rainwater collection systems can be built from reused materials. Homesteaders maximize water supplies without sacrificing fiscal responsibility by utilizing rainfall. This practice reduces their dependence on external water supplies and fosters a sense of independence in managing their resources.

For homesteaders who raise animals, livestock housing is an essential factor. When building shelters, it's crucial to use affordable and long-lasting materials to give animals a safe and comfortable space without breaking the bank. Using used or recycled resources instead of buying new building supplies is economical. These materials include pallets, reclaimed wood, and leftover corrugated metal. Craftiness and do-it-yourself abilities are essential for turning salvaged materials into livable and beautiful homes. Homesteaders can build livestock housing that satisfies their dedication to sustainable methods and financial limits by embracing creativity and inventiveness.

Bare-root trees are affordable for homesteaders who want to plant windbreaks or produce orchards. Compared to their potted counterparts, bare-root trees are usually less expensive because they are offered without soil or containers. They also cost less to ship because they are lighter and more straightforward. It is more likely that trees planted bare-root during the dormant season will be successfully established.

Homesteaders can obtain a range of fruit and nut trees without incurring significant costs thanks to the availability of bare-root trees at cheap prices from numerous nurseries and conservation programs. Implementing bare-root trees is in line with ecologically responsible methods and financial concerns.

A practical and affordable alternative, solar power is becoming increasingly popular among homesteaders searching for sustainable energy sources. Recent years have seen a sharp decline in the price of solar panels, opening up solar energy systems to homesteaders on a tight budget. Furthermore, government incentives, tax credits, and rebate programs may further lower the initial cost of solar system installations. Homesteaders can begin modestly by installing solar-powered irrigation pumps or portable chargers for small appliances and then progressively increase their solar infrastructure to meet their changing demands and financial resources. Gradually incorporating solar power fits homesteads' budgetary limits and puts them on the path to energy independence.

For homesteaders in colder climates, cost-effective and efficient heating options are essential. Rocket mass heaters offer a creative and affordable substitute for conventional wood burners. These heaters maximize heat output while using minimal amounts of wood fuel. For a tenth of the price of traditional wood-burning stoves, rocket mass heaters may be made using common materials like bricks, cob, and metal barrels. Because of their effectiveness in burning wood at high temperatures, less wood is consumed, which lowers costs and promotes sustainable forestry practices. Homesteaders can acquire warmth and comfort in their dwellings while upholding the values of affordability and environmental responsibility by implementing rocket mass heaters.

Vermiculture, or worm composting, is a waste management technique that is affordable and space-efficient for handling organic and kitchen waste. Red wiggler worms, which feed on organic materials, can be found in a simple worm bin made from wooden boxes or recycled containers. These hardworking worms break down kitchen scraps into nutrient-rich vermicompost, a great way to improve the soil. For homesteaders with limited resources, vermiculture is a tremendous waste management alternative because it requires little space and is inexpensive to set up. Vermiculture improves soil fertility and health in addition to handling waste disposal.

CHAPTER IV

Building Your Garden

Raised Beds vs. In-Ground Planting

When setting up their cultivating sites, every homesteader and gardener must choose between raised beds and in-ground planting. Every technique has advantages, disadvantages, and factors that affect the plants' general health and yield and the garden's visual attractiveness. This section explores the subtle differences between in-ground planting and raised beds, highlighting each method's advantages, disadvantages, and potential drawbacks to help homesteaders make decisions specific to their needs and tastes.

Homesteaders and gardeners are increasingly adopting raised beds because of their well-defined borders and raised growing surfaces. These beds are usually built out of wood, concrete blocks, or repurposed materials to provide enclosed planting areas above ground. The enhanced control over soil quality is one of the main benefits of raised beds. Raised beds allow homesteaders to customize the soil's composition for the best possible fertility, drainage, and structure. Homesteaders can establish the perfect growing conditions for plants with specialized soil requirements thanks to this level of personalization.

One of the main reasons raised beds are so popular is because of their improved drainage. The raised design allows excess water to drain away more effectively, avoiding soggy soil, a common issue in places with heavy rain. In addition to encouraging robust root systems, this enhanced drainage reduces the possibility of root rot and other water-related problems. Raised beds also allow for early spring planting since their raised design enables the soil to warm up faster than in-ground soil, prolonging some crops' growing season.

Raised beds offer containment, which helps with weed control. With clearly defined borders, erecting physical barriers against invasive plants is more straightforward, which lessens nutrient competition and reduces the need for heavy weeding. Furthermore, compared to compacted in-ground soil, the loose and friable soil in raised beds is less favorable to weed growth. This feature makes upkeep easier and makes the garden area look more aesthetically beautiful and well- organized.

Another essential advantage of raised beds is

accessibility, especially for homesteaders with mobility or physical restrictions. The raised design minimizes bending or kneeling by enabling comfortable waist-height gardening. This feature makes gardening more enjoyable, inclusive, and accessible to people with different physical abilities. Raised beds with features like trellises can help homesteaders contemplate long-term cultivation techniques by boosting the garden's efficiency overall and making tending to climbing plants easier.

Although raised beds have many benefits, there are drawbacks as well. The initial expense of building raised beds can be more than preparing an in-ground planting location, mainly if sturdy materials like cedar or composite lumber are used. For homesteaders on a limited budget, the cost of constructing raised beds may be a barrier; thus, carefully weighing the long-term advantages and return on investment is essential.

Raised beds require careful consideration of watering

techniques in addition to financial considerations. In addition to minimizing waterlogging, better drainage helps hasten soil drying, particularly in arid conditions. To ensure that plants get enough water, homesteaders must carefully monitor moisture levels and use strategic watering techniques. Raised beds take more upkeep overall, including installing drip irrigation systems, soaker hoses, or frequent hand watering.

Unlike raised beds, which are planned and contained, in-ground planting is done directly in the natural soil. Crops sown directly into the ground in broader open spaces are typical characteristics of this traditional method. The main benefit of in-ground planting is that it is inexpensive. Homesteaders only need to spend a little on soil preparation because they may work with the current land and make necessary amendments. Because of this, in-ground planting is a desirable choice for people who want to create extensive gardens without making a substantial initial investment.

The in-ground planting area's natural soil composition and structure support the general health of the plants. Plants in in-ground beds gradually grow extensive root systems that allow them to access natural water and fertilizer sources in the soil. Physical obstacles do not prevent plants' roots from growing freely, which helps them build a solid base and more effectively use nutrients. Therefore, in-ground planting encourages a symbiotic link between plants and the soil ecology around them.

The ability of in-ground planting to retain water is frequently praised. When appropriately modified with organic content, the native soil may successfully hold onto moisture. This can be especially helpful in areas with inconsistent rainfall or scarce water supplies. Because in-ground soil can hold moisture for extended periods, it requires less irrigation, which helps conserve water and supports sustainable gardening techniques.

Although in-ground planting has many advantages, there are specific issues with the structure and quality of the soil. The natural soil in some places could be compacted, poorly drained, or deficient in vital nutrients. To improve fertility and structure, homesteaders who plant in the ground must put a lot of time and effort into preparing the soil. This includes adding amendments like compost, organic matter, and soil conditioners. Several seasons may pass before the requirement for improved soil is met, necessitating persistence and patience in cultivating healthy soil.

Another thing to think about when planting in the ground is weed control. It is simpler for weeds to spread and compete with cultivated plants for resources without physical impediments. When implementing in-ground planting, homesteaders must be careful to control weeds. This can be done by mulching, weeding frequently, or employing cover crops to prevent weed growth. Although these methods work well, compared to the more limited setting of raised beds, they require extra maintenance.

One possible disadvantage of in-ground planting is accessibility, especially for those with physical impairments. Some homesteaders may find it challenging to bend, kneel, or perform tasks that need to be near the ground. Raised beds or container gardening might offer a more comfortable and convenient gardening experience in these circumstances.

The final decision between in-ground planting and raised beds is based on several variables, such as the homesteader's preferences, aims, and available resources. A hybrid strategy that incorporates both techniques might occasionally provide a well-rounded answer. Raised beds, for instance, can be used to cultivate particular crops intensively, while in-ground planting sites can support bigger or more extensive plantings. With this hybrid approach, homesteaders may take advantage of both approaches and customize their gardening plan to meet the particular requirements of their climate and terrain.

Companion Planting Strategies

An age-old agricultural technique called companion planting is carefully arranging plants adjacent to one another to promote reciprocal advantages. This symbiotic gardening method aims to establish a harmonic ecology where each plant contributes to the well-being of its neighbors, going beyond typical considerations of soil quality and sunlight. This section delves into the fundamentals, advantages, and tactics of companion planting, examining the complex web of interrelationships between plants and their potential to affect a garden's general health significantly.

Using a plant's inherent defenses to ward off pests and improve resilience to disease is one of the core ideas of companion planting. Numerous plants generate substances, such as chemicals or essential oils, that serve as organic deterrents for particular pests. Homesteaders can build a defense system that reduces the need for chemical interventions by carefully matching plants that enhance each other's ability to repel pests. To encourage a better and more resilient crop, growing fragrant herbs like basil or rosemary next to tomatoes will help ward off pests that frequently damage tomato plants.

The idea of trap cropping is a commonly used companion planting method in the field of pest control. Certain pests are drawn to particular plants, which deters them from attacking more valuable crops. For example, placing nasturtiums close to brassicas can serve as a sacrifice for cabbage worms since worms are attracted to nasturtiums but stay away from the brassica family. Using a tailored approach benefits the entire garden ecology by protecting the main crop and taking advantage of the inherent preferences of the pests.

Relationships in which plants share nutrients are also included in complementary planting. Strategic pairings can maximize nutrient consumption because different crops have different nutrient requirements or uptake processes. Because legumes can fix nitrogen from the air, they work well with crops that require nitrogen, such as corn or leafy greens. The legumes' nitrogen in the soil makes it available to nearby plants, establishing a natural and sustainable fertilization system. This interaction of nutrient-sharing enhances an ecology in gardens that is more balanced and self-sufficient.

Companion planting includes encouraging pollination, drawing beneficial insects, improving nutrient availability, and discouraging pests. Throughout the growing season, planting various blooming plants will attract pollinators like bees and butterflies. They are fruiting crops like tomatoes, peppers, and squash, which benefit from the presence of these pollinators since it increases the chance of successful pollination and fruit development. Furthermore, some companion plants release volatile substances that draw helpful insects like predatory beetles and ladybugs, which actively hunt down garden pests. Using an integrated approach to pest management, the garden ecosystem's natural equilibrium is promoted, and ecological principles are adhered to.

The "Three Sisters" method used in Native American agricultural practices is a famous illustration of companion planting. Squash, beans, and corn are cultivated in a symbiotic relationship where each plant offers certain advantages. Less trellising is required since the corn gives the beans a vertical structure to climb. In exchange, the nutrient-hungry corn benefits from the beans' ability to fix nitrogen in the soil. The broad leaves of the squash produce a natural mulch that inhibits the growth of weeds and helps to preserve soil moisture. The three crops' interdependence exemplifies the complex dance of companion planting, in which every plant serves to nourish the others.

Certain plant combinations can produce biological chemicals that impede the growth of nearby plants, known as allelopathic effects, in contrast to the harmonious interactions promoted by companion planting. For instance, juglone, a chemical molecule released by walnut trees, can impede the growth of numerous plants in their root zone. Companion planting requires understanding these allelopathic interactions since improper pairings can have unexpected results. Homesteaders should be aware of these interactions and choose companion plants that enhance rather than interfere.

Companion planting also includes spatial use, the deliberate arrangement of plants with different growth patterns and structures to make the most available sunlight and space. More sensitive crops can benefit from the shade that taller plants, like maize or sunflowers, can give them during the summer's heat. On the other hand, low-growing ground coverings, such as clover or creeping thyme, produce live mulch, keeping the soil moist and controlling weeds while providing a helpful understory for taller plants.

Companion planting adds ecological advantages and visual attractiveness to the garden. Landscapes are visually dynamic and diversified when colors, textures, and forms are purposefully mixed. Companion planting is beautiful because it creates a garden that is a tapestry of life through the creative arrangement of plants, which goes beyond the practical benefits of pest control and nutrient exchange. In addition to improving the entire garden experience, this aesthetic harmony encourages attention and a sense of connection with the natural world.

Companion planting effectiveness depends on carefully considering plant connections, including growth patterns, nutrient needs, and interactions with pests. Crop rotation, a technique closely related to companion planting, reduces the danger of soil-borne illnesses and nutrient depletion by preventing plants with comparable nutrient requirements from being continuously planted in the same spot. Effective companion planting requires careful planning, observation, and adaptation; homesteaders must keenly understand the dynamics inside their garden environment.

Although companion planting has many advantages, it is essential to understand that there is no one-size-fits-all approach to gardening. Understanding the particular requirements and traits of each plant and the specific circumstances of the surrounding environment is essential for successful implementation. A vital component of the process is experimentation and adaptation, which enables homesteaders to adjust companion planting tactics to the changing needs of their gardens.

To sum up, companion planting is evidence of the complex network of interrelationships in the garden ecosystem. Beyond the valuable benefits of improved pollination, nutrient sharing, and pest control, companion planting is an all-encompassing gardening method that honors all living things' interdependence. The garden develops into a thriving community where plants encourage and enhance one another and serve as a place for cultivation. Homesteaders go on a journey of co-creation with nature as they master the delicate dance of companion planting, cultivating a robust and peaceful garden that embodies the wisdom of tried-and-true ecological principles.

Tips for Maximizing Garden Space

When it comes to homesteading and gardening, the valuable space in a garden needs to be planned carefully to maximize its usage and yield. Optimizing garden space requires careful planning, effective management, and innovative thinking for a small urban plot, a backyard garden, or a larger homestead. This section examines various techniques and strategies to help homesteaders maximize their garden area and cultivate abundance within the limitations of available square footage.

Effective plan design is an essential component of making the most of garden space. A thoughtful garden design considers elements like wind patterns, sunshine exposure, and the particular requirements of each plant. The growth area can be significantly increased using vertical space through trellising or vertical gardening. Training plants to grow upward against a wall, fence, or specialized trellis is known as vertical gardening. Vegetables that grow well on vines, like cucumbers, beans, and peas, are excellent candidates for vertical gardening, which enables homesteaders to make the most of the space above the ground that is sometimes underutilized.

Another effective layout technique is square foot gardening, which Mel Bartholomew popularized. This strategy divides the garden into small, manageable square-foot pieces, each devoted to a particular plant or crop. This method makes the most available area and makes planting, caring for, and harvesting easier. Crops are arranged in defined squares to avoid wasted space between rows and to create a dense, productive garden plan.

The companion planting technique, which is closely related to making the most of garden space, matches plants in a way that will complement one another. Plants can have synergistic connections in which one plant helps to resist pests that affect its companion or promote the growth of nearby crops. Homesteaders can create a more robust and healthy garden ecology while optimizing their available area by interplanting compatible species.

Raised beds are a flexible way to make the most of garden space, particularly in places with compacted soil or poor quality. Raised beds maximize available space with well-defined borders and controlled drainage and soil quality environment. The square or rectangular design of raised beds makes arranging and dividing crops simple. Raised beds can make gardening more accessible by reducing the need for excessive bending or kneeling, making gardening more accessible and ergonomic.

A dynamic method of managing gardens, succession

planting entails sowing fresh crops as soon as one harvest is completed. By maintaining a constant planting cycle, garden space is constantly utilized, extending the growth season and increasing yield. Homesteaders can ensure a consistent supply of fresh produce throughout the growing season by scheduling subsequent plantings according to various crops' maturity and harvest seasons.

Using pots and containers is a sensible choice for people with difficult soil conditions or a little garden area. Homesteaders can grow various plants in containers on their patios, balconies, or even windowsills. Because they are made of different materials, sizes, and shapes, containers provide flexibility in layout and design. In pots, herbs, salad greens, and small fruit kinds grow well and expand the garden into areas that might not be used for gardening.

Beyond trellises, vertical gardening can involve tiered constructions, wall-mounted pots, and hanging baskets. Plant containers or pockets can be added to walls and fences to create vertical gardens. Hanging baskets give trailing plants more room to flourish when suspended from pergolas or above frames. Like staircases, tier planters make the most of vertical space and create an eye-catching garden structure. These vertical planting methods provide the garden with an attractive touch in addition to making effective use of available space.

Homesteaders can produce crops past customary

harvest dates by extending the growing season with cold frames, hoop houses, or tiny greenhouse structures. By shielding plants from wind, frost, and temperature fluctuations, these season-extending structures foster a microclimate ideal for year-round gardening. Cold frames are translucent lidded structures that are usually low to protect plants sensitive to cold or for early seed initiation. Constructed with arching frames coated with plastic or row cover material, hoop houses offer a cost-effective and adaptable way to prolong the growth season. Small greenhouses with more permanent structures provide heat-loving crops with a regulated atmosphere so that homesteaders can try a wider variety of plants.

By producing several crops in one location, intercropping techniques maximize available space. Companion plants with varied growth patterns, root systems, and maturation dates can be interplanted to create a harmonious and space-efficient landscape. For example, crops that increase, like lettuce or radishes, might be put in between rows of plants that grow more slowly, like tomatoes or peppers. In addition to making the most of available space, intercropping increases overall garden resilience encourage biodiversity and lessens insect burden.

Raised beds are a flexible way to make the most of garden space, particularly in places with compacted soil or poor quality. Raised beds maximize available space with well-defined borders and controlled drainage and soil quality environment. The square or rectangular design of raised beds makes arranging and dividing crops simple. Raised beds can make gardening more accessible by reducing the need for excessive bending or kneeling, making gardening more accessible and ergonomic.

A dynamic method of managing gardens, succession planting entails sowing fresh crops as soon as one harvest is completed. By maintaining a constant planting cycle, garden space is constantly utilized, extending the growth season and increasing yield. Homesteaders can ensure a consistent supply of fresh produce throughout the growing season by scheduling subsequent plantings according to various crops' maturity and harvest seasons.

Using pots and containers is a sensible choice for people with difficult soil conditions or a little garden area. Homesteaders can grow various plants in containers on their patios, balconies, or even windowsills. Because they are made of different materials, sizes, and shapes, containers provide flexibility in layout and design. In pots, herbs, salad greens, and small fruit kinds grow well and expand the garden into areas that might not be used for gardening.

Beyond trellises, vertical gardening can involve tiered constructions, wall-mounted pots, and hanging baskets. Plant containers or pockets can be added to walls and fences to create vertical gardens. Hanging baskets give trailing plants more room to flourish when suspended from pergolas or above frames. Like staircases, tier planters make the most of vertical space and create an eye-catching garden structure. These vertical planting methods provide the garden with an attractive touch in addition to making effective use of available space.

CHAPTER V

Raising Happy and Healthy Livestock

Choosing the Right Animals for Your Space

Bringing animals onto a homestead is a crucial step toward sustainability and self-sufficiency. Whether you're driven by a need for dairy, meat, or eggs or help managing your land, choosing the appropriate animals for your area is essential. The best livestock depends mainly on your goals, the size of your property, and the resources at your disposal. This section explores the key elements and strategic criteria that help homesteaders select the correct animals, promoting a positive and fruitful coexistence between people and their animal companions.

Making a detailed assessment of your available space is the first step in choosing the ideal animals for your farm. While individuals with little area find poultry or small livestock like rabbits more feasible, homesteaders with vast acreage can support larger livestock like cattle, goats, or pigs. The kind of animals you can keep, and the infrastructure needed to house and care for them are both influenced by the size and design of your property. Sufficient room allows animals to behave as they naturally would, which lowers stress and improves general wellbeing.

It is crucial to consider local zoning laws before bringing animals onto your homestead. Zoning regulations may include limitations on the kinds and quantity of animals permitted and the buildings that can be used to house them. Acquainting yourself with these regulations guarantees adherence and averts possible legal complications. Furthermore, speaking with nearby residents or consulting the community's guidelines can provide you with any limitations or concerns regarding noise, smell, or other aspects that might influence your chosen animals.

Making wise judgments requires knowing why you are bringing animals onto your ranch. Numerous animals fulfill various purposes, such as supplying food and fiber, providing company, or aiding with land management techniques. Goats or cows are good options for individuals who want to produce dairy products, while hens are a good choice for those who want a consistent supply of fresh eggs. With their ability to feed on weeds and insects, animals like ducks and geese can aid in the control of pests. A meaningful and mutually beneficial relationship between homesteaders and livestock is ensured by carefully matching your objectives with each species' unique traits and qualities.

Evaluating the resources available for animal care is essential to figuring out whether bringing a particular species to your homestead is feasible. An adequate food, water, shelter, and veterinary care supply is necessary for ethical animal husbandry. Because every animal is different, preparing for their dietary requirements is essential. Furthermore, a practical and long-lasting approach to homestead animal management is ensured by considering the time and effort needed for routine maintenance, care, and crises.

Their temperament and conduct mainly determine the compatibility of potential livestock with your property and lifestyle. Certain animals, like goats, need careful supervision and safe fencing because of their curiosity and naughty reputation. On the other hand, guinea pigs and rabbits might be more tame and able to adapt to smaller living areas. Comprehending the distinct behavioral characteristics of various species enables homesteaders to foresee obstacles and execute efficient management tactics to guarantee the welfare of both fauna and domestic animals.

The temperature and environmental factors in your area affect whether livestock are suitable. Animals acclimating to a particular climate are more resilient and require less care. For example, livestock or fowl varieties adapted to colder areas may struggle to thrive in hot, muggy weather. The selection of animals that can resist local weather patterns and contribute to a resilient and prosperous homestead is guided by assessing your area's climate-related opportunities and difficulties.

Long-term planning requires considering prospective livestock's lifetime and reproductive traits. Certain animals, such as chickens, have quick cycles of reproduction and relatively short lifespans, which means they can produce an abundant supply of meat or eggs. Larger animals, like horses or cows, on the other hand, reproduce more slowly and take longer to gestate. A holistic approach to animal husbandry is ensured by understanding the commitment required in providing care for animals throughout their life stages, including breeding, birthing, and potential end-of-life issues.

The influence of animals on the land must be carefully considered before integrating them into the larger homestead ecosystem. Grazing cattle in different pasture areas on a rotational basis helps prevent overgrazing and encourages healthy vegetation. Furthermore, considering animals' capacity to cycle nutrients—for example, applying chicken dung as fertilizer to garden beds—contributes to a closed-loop system in which waste is turned into a helpful resource. Regenerative agriculture is based on the fundamental idea of animals and land working together to create a balanced and regenerative homestead environment.

A resilient and multipurpose homestead might benefit from having a diverse livestock herd. Combining species with complementary traits improves the homestead ecosystem's productivity and efficiency. For instance, more extensive cattle may aid with soil aeration or clearing areas, while chickens can help manage insect problems in the garden. By including animals with various functions, a dynamic and interconnected system is created in which every species has a distinct role to play in the larger story of the farm.

Homesteaders should base their decisions primarily on ethical and compassionate treatment of animals, regardless of practical factors. Giving animals access to clean water and a safe and comfortable place to live is morally required, as a healthy diet. A dedication to moral animal husbandry methods is demonstrated by the attention paid to their mental and physical wellbeing, which includes providing opportunities for socialization and natural activities. Building a respectful and loving relationship with your animals improves their quality of life and makes homesteading more rewarding and significant.

Selecting the ideal animals for your property is a dynamic process that requires constant assessment and modification. Homesteaders may find new possibilities, difficulties, and objectives as they gain experience, which affects their decision-making. Consulting with seasoned homesteaders, nearby agricultural extension offices, or neighborhood associations can yield insightful advice and helpful connections. Through a combination of pragmatic factors, moral precepts, and in-depth knowledge of the particular circumstances surrounding the homestead, homesteaders set the stage for a peaceful and mutually beneficial relationship with their animal friends when choosing livestock.

Building Coops and Shelters

Building coops and shelters, which give cattle a safe and comfortable place to reside, is a crucial part of managing a homestead successfully. The planning and building these buildings is essential to guaranteeing the welfare of the occupants, be they goats, chickens, rabbits, or other animals. This section examines the critical factors, design tenets, and helpful advice for creating coops and shelters that support a robust and healthy homestead ecology and provide necessities for cattle.

Knowing the particular requirements of the intended occupants is one of the most important factors to consider while constructing coops and shelters. The needs of various livestock species vary concerning temperature, ventilation, space, and security. Goats, for instance, could need raised platforms and enough room for social activities, while chickens need roosting bars and nesting boxes in their coop. Adapting the design to the distinct qualities of the animals guarantees that the structure fulfills its intended function.

Placing coops and shelters on the homestead is an important choice that affects the animals' welfare and the property's general usability. Buildings on high ground with adequate drainage prevent waterlogging and guarantee a dry and cozy living space. Considering the predominant wind patterns to offer protection from severe weather is imperative. To enable effective waste management and resource utilization, proximity to other components of the homestead, such as gardens or compost areas, should also be considered.

One of the most important aspects of giving the animals enough living space is the size of the coop or shelter. Overcrowded livestock may experience stress, illness, and violent behavior. Larger animals like goats can need more space, but as a general rule of thumb, each chicken should have two to three square feet. An ample room encourages social relationships, lets animals exhibit their natural habits, and is suitable for general health and wellbeing.

Ventilation is crucial in controlling moisture, temperature, and air quality when designing coops and shelters. Sufficient ventilation promotes a healthier animal atmosphere by dispersing airborne pathogens, humidity, and ammonia. During the colder months, windows, vents, or open-air designs allow natural ventilation while providing draft protection. In addition to making livestock more comfortable, proper ventilation reduces the possibility of respiratory problems and mold growth inside the building.

Because insulation is a barrier against heat and cold, it is especially crucial in areas with harsh weather. The animals live in a more stable and comfortable habitat in well-insulated coops because they stay cool in the summer and retain warmth in the winter. Straw, hay, or foam boards are insulating materials used in the roof and walls to improve energy efficiency and control temperature. Homesteaders can minimize their dependency on external energy sources and maximize their ability to adjust the temperature by carefully placing windows or ventilation openings to harness the power of nature.

Predator protection is a top priority while building coops and shelters. Ensuring the protection of cattle requires guarding the structure against potential dangers, such as foxes, raccoons, or birds of prey. To prevent unwanted access, windows, doors, and other openings should be equipped with predator-resistant materials, hardware cloth, or reinforced wire mesh. Predators that burrow can be discouraged by erecting underground barriers, including burying wire mesh around the perimeter. Keeping doors closed and apertures secured at night adds even more security to the coop.

Coops and shelters' longevity and environmental impact are primarily determined by the materials used during construction. The resistance of pressure-treated timber, redwood, and cedar to insects and deterioration makes them popular alternatives. Reclaimed or recycled materials can be utilized to repurpose things like rescued timber or pallets in an environmentally responsible manner. Metal roofing ensures a lengthy lifespan for the shelter by offering resilience and resistance to the weather. Additionally, the general health of the animals is improved by using safe and non-toxic materials for interior finishes like paints and sealants.

When thoughtful design and arrangement are used, coops and shelters are more functional and efficient. Cleaning and maintenance activities are made more accessible using sliding doors, easy-access nesting boxes, and retractable litter trays. The homesteader's convenience is increased by the structure's ample storage for supplies, tools, and feed. Including architectural features that encourage natural behaviors—like platforms for goats or perches for chickens—increases the animals' general happiness and wellbeing.

Accessibility for homesteaders to enter, clean, and oversee the coop is a critical but frequently disregarded architectural element. Wide enough doors are essential for simple access during regular duties like feeding and cleaning in coops. All regions of the coop should be easily accessible thanks to a well-planned arrangement, which lessens stress for the animals and the homesteader. A more pleasurable and helpful household experience can be achieved by considering ergonomic design elements, such as comfortable working heights and practical storage.

Designing coops and shelters requires careful consideration of water management, especially in areas where weather patterns are erratic. Waterlogging and possible flooding are avoided during high rainfall by maintaining adequate drainage around the building. Installing gutters and downspouts will help reduce erosion risk and keep the area dry by diverting precipitation from the chicken coop. On the homestead, gathering rainwater for irrigation systems is another way to support sustainable water management techniques.

Designing coops and shelters with sustainable and energy-efficient elements aligns with the ideals of environmentally aware homesteading. Incorporating solar panels can reduce reliance on external energy sources by powering heating systems or providing lighting. By strategically placing windows or skylights, natural light can be maximized, reducing the need for artificial lighting during the day. A closed-loop water management system can be fostered by integrating rainwater harvesting technologies to gather and store water for garden irrigation or animal consumption.

The visual component of shelter and coop design gives homesteading a creative edge. Carefully considering the landscaping surrounding the buildings, like adding vines or shrubs, improves their aesthetic appeal while offering wind and shade protection. A unified and visually appealing environment is achieved by incorporating design features that complement the homestead's overall aesthetics, such as matching the coop to the house's or the surrounding structures' architectural style.

Feeding and Health Considerations

A homestead's overall prosperity and sustainability largely depend on the essential duties of feeding and caring for its cattle. Whether an animal produces milk, eggs, wool, meat, or other goods, its well-being directly affects these factors, as well as its longevity and productivity. To build resilience, avoid disease, and encourage the flourishing of animals within the homestead ecosystem, this section focuses on behaviors that are critical to consider when feeding and caring for livestock on homesteads.

Ensuring each species receives a balanced and nutrient-rich diet is essential to successful animal management. Developing proper feeding practices for cattle requires a thorough understanding of their nutritional needs. Certain animals have different dietary requirements depending on age, species, reproductive status, and general health. For example, growing chicks, nursing goats, and aged rabbits all need different nutrients. To ensure the best possible health and production, homesteaders should take the time to investigate and create diets that satisfy these various requirements.

Providing premium forage and feed is the cornerstone of animal nutrition. Pasture management is a crucial factor for homesteaders who have access to grazing grounds. Grazing livestock in rotation—moving them to various pasture parts regularly—avoids overgrazing, promotes regenerative land management, and guarantees a steady supply of fresh fodder. Good hay or other feeds can be added to pasture to maintain nutritional balance and prevent nutrient deficits, mainly when forage is scarce.

When pasture access is restricted or nonexistent,

choosing commercially available feeds carefully becomes crucial. Superior, species-specific diets devoid of impurities or additives promote animals' general health and welfare. A thorough understanding of the protein, fiber, and mineral levels found in commercial feeds helps one make decisions that align with each species' unique requirements. A varied and nutrient-rich diet is produced by balancing commercial feeds with fresh forage and additional sources, like leftover kitchen scraps or garden waste.

Water is essential to the health and nutrition of animals. For animals' welfare, they must have access to pure water. It's crucial to regularly check on water sources and clean water containers and make sure you're getting enough water. In colder climates, keeping water from freezing becomes a winter difficulty that requires heated water or other techniques to guarantee a steady water flow.

Homesteaders should consider how their eating methods will affect their long-term health and address their fundamental nutritional needs. Inadequate or excessive feeding can result in obesity, malnutrition, and metabolic abnormalities, among other health problems. Homesteaders can modify feeding schedules on their animals' body condition, weight, and general health by regularly monitoring these factors. Furthermore, watching animals eat offers essential insights regarding their behavior, hunger, and any possible symptoms of illness or suffering.

Supplementation is frequently required to sustain the health of cattle or address specific nutritional deficits. For instance, giving out mineral supplements that include vital components like calcium, phosphorus, and selenium can stop inadequacies that could cause immune system damage, skeletal problems, or poor reproductive outcomes. To perform routine assessments and choose the right supplements based on the particular requirements of their animals and the local soil conditions, homesteaders should speak with veterinarians or livestock nutritionists.

While diet is a critical component of animal health, preventative care is just as essential. A regular healthcare program that includes immunizations, parasite management, and general wellness checks must be established to prevent infections and treat health difficulties in their early stages. Vaccinations shield animals from common illnesses that are endemic in the area, and parasite management techniques like deworming aid in preventing infestations that can cause malnourishment and general debilitation.

Routine veterinary consultations and developing a solid rapport with an experienced veterinarian aid effective healthcare management. Veterinarians can perform diagnostic procedures, give advice on diet and parasite control, and provide necessary assistance regarding immunization regimens. They are also essential in recognizing and treating new health problems, providing quick and efficient care that boosts the resilience of the homestead cattle.

Animal foot care is a topic that is frequently disregarded but needs to be addressed, especially when it comes to species like sheep and goats. Hoof overgrowth can cause lameness, pain, and increased vulnerability to infections. Regular hoof trimming promotes general well-being and guarantees that animals can move quickly. It is done using the right equipment and methods. Homesteaders should include regular clipping into their management procedures and become knowledgeable about their livestock species' unique foot care needs.

Preventing illness and fostering health depends heavily on keeping livestock's living quarters clean and hygienic. Frequent cleaning lowers the chance of bacterial or fungal diseases in coops, shelters, and bedding. In addition to promoting a clean living environment, effective waste management—which includes getting rid of soiled bedding and manure—also stops the growth of pests and parasites. Maximizing ventilation in enclosed places is essential to reduce humidity and prevent respiratory irritants from building up.

The proper disposal of manure is an integral part of effective waste management, which goes beyond the coop or shelter. Manure can help improve soil fertility in fields or gardens when properly composted. Composting closes the loop on nutrient cycling within the homestead ecosystem while simultaneously lessening the environmental impact of waste and promoting sustainable agriculture practices. As a natural fertilizer, composted manure encourages strong plant development and supports a regenerative method of land management.

Strategies for preventing disease also include methods for lessening animal stress. Reducing stresses, including sudden dietary changes, crowded living situations, and exposure to harsh weather conditions, can support a healthy immune system. Giving animals enough room, suitable shelter, and a peaceful atmosphere promotes their mental health and increases their resistance to illness in general.

Many homesteaders choose a holistic approach, incorporating natural healthcare techniques and herbal medicines into the overall healthcare strategy. Herbs with natural immune-boosting or anti-parasitic effects include garlic, oregano, and chamomile. Even though using herbal treatments on animals should be done carefully and preferably after consulting a veterinarian, many homesteaders find that adding these organic components to their animal care practices beneficial.

Animal mental and emotional health is included in the notion of holistic healthcare. Similar to people, livestock also have social needs and experience stress and anxiety. Their mental health is enhanced when social contact possibilities are provided, such as companionship for lonely animals or proper herd dynamics for gregarious species. A stimulating and engaging living environment is produced via enrichment activities, such as providing toys, scratching posts, or environmental stimulation.

In summary, providing for domestic cattle's nutritional needs and overall well-being are complex and interconnected facets of adequate animal husbandry. Homesteaders support their animals' resilience and well-being by emphasizing sustainable practices, preventive treatment, and balanced feeding. In addition to increasing livestock productivity, the dedication to proactive and comprehensive methods of feeding and healthcare also complies with the tenets of sustainable and regenerative homesteading. Homesteaders foster a robust and peaceful relationship between humans and their animal friends through regular veterinary care, clever management methods, and careful attention to each species' specific needs within the homestead environment's dynamic tapestry.

CHAPTER VI

Sustainable Practices

Composting and Recycling

Sustainable homesteading is based on the principles of composting and recycling, which create a dynamic and regenerative loop that turns organic waste into valuable resources. In addition to lessening the environmental effects of domestic and agricultural activities, these methods also help to maintain healthy soil, preserve resources, and establish closed-loop systems. This section delves into the practice and theory of composting, its advantages for homestead ecology, and the idea of recycling concerning materials other than organic waste.

On the farm, composting is essential to sustainable waste management. Fundamentally, composting is the organic and garden waste that naturally breaks down into a nutrient-rich soil conditioner through decomposition. The homesteader's job in this process is to break down organic materials by fostering the growth of microorganisms such as bacteria, fungi, and insects. What could be regarded as waste is turned into a valuable resource for improving soil fertility and structure by a well-maintained compost pile.

Understanding the elements that comprise a balanced compost pile is the first step in building a successful composting system. Green components, such as manure, fresh plant matter, and kitchen trash, combine moisture and nitrogen. Dried leaves, straw, and cardboard are examples of brown materials that give carbon and give the pile the structure it needs. Balancing these green and brown components ensures the proper carbon-to-nitrogen ratio, creating the perfect microbial activity environment.

Another critical factor in the compost pile's success is its physical composition. Regular turning of the compost adds oxygen, speeds up the decomposition process, and keeps the pile from going anaerobic, which could produce foul smells. Like that of a wrung-out sponge, sufficient moisture content encourages microbial activity without making the environment too wet. Additionally, placing the compost pile in a spot with plenty of sunlight and drainage speeds up decomposition. A cover also aids in keeping moisture and heat in the compost.

Composting has many advantages over waste reduction.

As a nutrient-rich soil conditioner, compost strengthens the structure of the soil, increases water retention, and promotes the growth of healthy microbial populations. Homesteaders help to create healthy, nutritious soil that supports vigorous plant growth and grows crop resilience to pests and diseases by adding compost to garden beds or agricultural fields. Composting is a closed-loop process that fits nicely with the regenerative ideas of sustainable agriculture, which returns organic matter into the soil to finish the nutrient cycle.

Composting is another helpful method for handling kitchen trash, which lowers the quantity of organic material disposed of in landfills. Food scraps, coffee grounds, and vegetable peelings in the compost pile find a new use and become an invaluable resource instead of releasing methane, a potent greenhouse gas released by the breakdown of organic materials in landfills. Homesteaders actively contribute to lessening the environmental impact of disposing of organic waste and their ecological footprint by directing kitchen garbage into the compost pile.

Recycling on the farm goes beyond composting and includes a broader range of items, such as Glass, plastic, paper, and metal. Recycling techniques assist in lowering the need for new raw materials, save energy, and lessen the adverse environmental effects of extracting and processing virgin resources. Homesteaders can ensure a systematic approach to trash diversion by setting up specific recycling containers for different materials.

Paper product recycling is critical when it comes to sustainable homesteading. Paper, a waste product frequently found in homes, can be recycled to create new paper goods, lessening the need for deforestation and the environmental damage that paper manufacturing causes. Homesteaders show their dedication to appropriate waste management and actively participate in resource conservation by gathering and recycling cardboard, newspapers, and other paper products.

Similarly, recycling metal and glass containers lowers the energy needed to create new materials. Glass is a sustainable packaging material since it can be recycled indefinitely without losing its quality. People who adopt a homesteading lifestyle by recycling metal and glass help reduce waste going to landfills and promote the ideas of a circular economy, which reuses and recycles commodities to reduce their environmental impact.

Plastic recycling is still a crucial part of responsible waste management, even if it can be difficult because there are many different forms of plastic, and certain things aren't very recyclable. Homesteaders should concentrate on using fewer single-use plastics, choosing reusable substitutes, and endorsing goods with little to no packaging. Sorting plastic waste into the correct categories when recycling is an option helps keep recyclable materials out of the trash and enables them to be recycled into new goods.

Yard trash can also be recycled and composted, including leaves, clipped branches, and grass clippings. Homesteaders can use these items to make wood chips, mulch, or compost instead of throwing them in the trash. Mulching retains soil moisture, inhibits weed growth, and enriches the soil by decomposing organic waste. Woody materials can also be chopped and added to a compost pile as a carbon-rich component or as a sustainable and natural mulch.

Recycling and composting are dynamic processes that must be incorporated into the homesteading way of life. Cultivating a responsible waste management approach requires learning about local recycling regulations, involving community recycling initiatives, and keeping up with emerging technologies. Composting bins, specific recycling spaces, and unambiguous labeling can also expedite the process and motivate family members to participate actively in sustainable activities.

A key component of encouraging composting and recycling habits is educational outreach inside and outside the homesteading community. Workshops, neighborhood gatherings, or internet forums offer chances for information exchange, idea sharing, and displaying effective homestead composting and recycling practices. Homesteaders support a more significant movement toward waste reduction and sustainable living by encouraging an environmentally conscious society.

Recycling and composting are related, which aligns with the ideas of permaculture, which views trash as an asset rather than a burden. The homestead ecosystem serves as a working example of how organic materials and matter can be recycled, with waste being converted into valuable resources that improve the production and general health of the land. This closed-loop system exemplifies homesteading's capacity for regeneration, as thoughtful actions result in constructing a robust and sustainable house.

To sum up, recycling and composting are essential habits for homesteaders who want to develop sustainability daily. Homesteaders positively impact the environment and the resilience of their homestead ecosystem by turning organic waste into nutrient-rich compost and removing recyclables from landfills. Regenerative agriculture, circular economies, and appropriate waste management embody these practices, representing a comprehensive approach to sustainable living. By recycling and composting garbage, homesteaders appropriately manage personal waste, positively influence the larger ecological landscape, and promote environmental stewardship in their communities.

Rainwater Harvesting

Rainwater harvesting is a long-standing and traditional method of utilizing precipitation to meet various water needs, encourage sustainability, and lessen the effects of water scarcity. For homesteaders, this old-fashioned method offers a cutting-edge, resource- and environmentally-conscious strategy for addressing the problems of managing water resources. This study delves into the fundamentals, advantages, and practical application of rainwater harvesting on the homestead, examining how this traditional method harmonizes with modern notions of environmental care and self- sufficiency.

Rainwater harvesting is gathering and storing rainwater for later use. Historically, societies have used this technique, from indigenous communities in dry locations to ancient Rome. Homesteaders are finding that rainwater harvesting is a viable way to ensure a decentralized and dependable water supply in the face of growing worries about water scarcity and the environmental effects of conventional water sources.

Distribution systems, storage tanks, conveyance systems, and catchment surfaces are the main parts of a rainwater collecting system. Rainwater can be effectively collected on catchment surfaces, usually the roofs of houses or other structures. A conveyance system then directs the rainwater through gutters and downspouts and into storage tanks. The gathered rainwater is stored in these concrete, metal, or plastic containers for later use. Distribution systems—consisting of pipes and pumps—deliver the collected rainwater to locations that require it, such as backyards, pastures, or even homes.

The capacity of rainwater collection to offer a decentralized and sustainable water source is one of its main benefits. Homesteaders can benefit from the natural abundance of rainfall instead of depending only on centralized water services, which will lessen the burden on groundwater resources and municipal water supplies. By promoting self-sufficiency, this decentralized strategy helps homesteaders become less reliant on outside water sources and more adaptable to interruptions or shortages of water.

Rainwater gathering has significant positive environmental effects and promotes water resource conservation and appropriate use. Homesteaders actively contribute to water conservation initiatives by avoiding runoff and preventing soil erosion through rainfall collection. This protects aquatic bodies from pollution brought on by runoff transporting pollutants from impervious surfaces, in addition to aiding in the preservation of regional ecosystems. Furthermore, by reducing the need for energy-intensive water purification procedures, rainwater harvesting lowers the carbon footprint associated with traditional water supply systems.

By maximizing the effectiveness of water distribution, rainwater gathering enables homesteaders to support sustainable water use. Harvested rainwater can be deliberately directed to places of need, such as gardens or orchards, where it functions as a vital irrigation source, by developing and executing distribution systems. This technique maximizes its use by strategically applying rainwater and ensures that every drop helps crops grow and thrive. It also reduces water loss, often linked to traditional irrigation techniques.

Rainwater collecting can be used for home and agricultural needs, demonstrating its versatility. Rainwater collection can be utilized for various tasks, including laundry, toilet flushing, and even delivering potable water after the necessary filtering and treatment. Even while rainwater needs to be treated appropriately before it can be used for most domestic purposes, using it for non-potable purposes eases the burden on municipal water supply and lowers homesteaders' water bills. In addition, incorporating rainwater into household water systems improves the homestead's overall resilience, particularly in times of water scarcity or limitations.

Implementing a rainwater harvesting system

necessitates careful consideration of several aspects, starting with catchment surface design. The kind and state of the roof significantly impact the quality of the collected rainwater. Roofs with surfaces like metal, tile, or asphalt shingles are ideal; however, roofs with moss or algae development might need to be cleaned before usage. Homesteaders should also consider the catchment area's size since larger surfaces produce more water, and the roof's slope affects how quickly and effectively rainfall runoff occurs.

Conveyance systems are essential in moving rainwater from collection surfaces to storage tanks. When installed and maintained correctly, gutters and downspouts guarantee effective water collection and save water waste. Before water enters storage tanks, filtration systems—such as screens or mesh filters—help remove impurities, debris, and leaves. Harvested rainwater is made even better by installing a first-flush diverter, which directs the first and dirtiest runoff away from the storage tanks.

Another important factor in constructing a rainwater

collecting system is the choice of storage tanks. Tanks are available in various sizes, forms, and materials, each with pros and cons. Polyethylene tanks are popular since they are inexpensive, lightweight, and corrosion-resistant. Although concrete tanks are durable, they can be more expensive and difficult to install. Although metal tanks are strong and don't produce algae, they can rust with time. Various aspects, including the planned use of the captured rainwater, the local climate, and the budget, influence the choice of tank material.

To guarantee the quality of collected rainwater, storage tank integrity, and cleanliness must be maintained. Frequent maintenance, cleaning, and inspection stop the accumulation of pollutants, algae, or sediment that can make the water unfit for use in different ways. Well-crafted tank covers and lids minimize the chance of algae growth and preserve water quality by shielding the stored water from sunlight.

Distribution systems must be created with the homestead's unique requirements in mind. Gravity-fed systems, which use the power of gravity to move water from storage tanks to designated areas, are straightforward and low-energy. They need to be carefully planned to guarantee appropriate slope and elevation. Pump-based systems can pressurize and distribute collected rainwater in scenarios where gravity-fed systems are unfeasible, particularly in regions with diverse topography. Various elements, including topography, elevation variations, and the water's planned uses, influence the decision between pump-based and gravity-fed systems.

It is necessary to take extra precautions to guarantee the safety of collected rainwater for drinking purposes. Rainwater may include pollutants that make it unfit for drinking without proper treatment, even if it is widely regarded as acceptable for non-potable applications like washing or irrigation. To improve the quality of rainwater for drinking, homesteaders can utilize filtration and purification techniques such as carbon filters, sediment filters, and ultraviolet (UV) disinfection. Ensuring that harvested rainwater satisfies the requirements for safe consumption can be achieved by testing it regularly or consulting with specialists in water quality.

Rainwater harvesting has advantages beyond a single farmhouse and enhancing environmental and community resilience. The widespread use of rainwater harvesting techniques might lessen the strain on current water supplies and alleviate the effects of water shortages in areas that experience recurrent droughts or scarcity of water. Rainwater harvesting becomes a scalable and community-focused method of managing water resources as more and more communities realize the benefits of decentralized and sustainable water solutions.

Promoting the broad use of rainwater collection requires educating the community about its advantages and practical applications. Workshops, neighborhood gatherings, or cooperative projects offer chances to exchange expertise, show off effective rainwater collection systems, and promote a water-saving mindset. Homesteaders support a more significant movement toward sustainable water practices by providing people and communities with the knowledge and resources to set up rainwater gatherings.

In summary, rainwater collecting is a comprehensive and environmentally friendly method of managing water on the property. The use of precipitation by homesteaders lessens their dependency on Conserve resources, maintain consolidated water supplies, and support environmental preservation. This age-old method offers a flexible and decentralized response to the problems associated with water scarcity by fusing with contemporary technology and self-sufficiency concepts. By adopting rainwater collection, homesteaders ensure a steady water supply for their needs and significantly contribute to building a more resilient and sustainable future for their communities and themselves.

Implementing Permaculture Principles

The holistic design concept of permaculture, based on ecological sustainability, biodiversity, and resilience, offers a guide for building self-sufficient and regenerative homesteads. Derived from the terms "permanent" and "agriculture," permaculture encompasses a wide range of systems that are in harmony with natural ecosystems and go beyond conventional farming methods. To build flourishing, resilient, and sustainable landscapes, homesteaders can use the fundamental ideas of permaculture, which we will examine in this section.

An understanding of and relationship with the natural world is the fundamental tenet of permaculture. Permaculturists thoroughly examine the land, its climate, terrain, water patterns, and existing ecosystems before implementing any design. Homesteaders are better able to comprehend the distinctive qualities of their property and choose design features that will blend in with the environment thanks to this in-depth observation. Homesteaders lay the groundwork for developing regenerative and environmentally balanced systems by developing a strong bond with the land.

Designing with an emphasis on efficiently capturing and using energy is one of the fundamental principles of permaculture. This theory aims to maximize the absorption and utilization of sunlight, water, and wind while acknowledging the significance of energy flows in natural systems. Homesteaders can maximize sunshine exposure for solar panels and gardens by carefully positioning buildings and plants to capture solar energy. While wind turbines and windbreaks are placed deliberately to harness the force of the wind, rainwater harvesting and the implementation of water catchment systems harness the power inherent in precipitation. Permaculturists develop efficient and sustainable household systems by intentionally including these energy sources in their plans.

At the heart of permaculture design is the idea of yield. This idea highlights how crucial it is to design systems that offer a variety of advantages and purposes. Every design component should have more than one function to ensure that every facet of the homestead enhances overall productivity. Fruit trees, for instance, produce an abundance of mouthwatering fruits in addition to offering shade, promoting biodiversity, and supplying organic matter for the soil. Including livestock in the system yields meat, milk, or eggs, improves soil fertility through manure, controls vegetation, and encourages organic pest control. The goal of achieving numerous yields produces resilient, multipurpose landscapes that maximize the utilization of the resources at hand.

One of the central tenets of permaculture is the efficient use of resources, which emphasizes the significance of reducing waste and maximizing resource cycles. Homesteaders can establish self-sustaining ecosystems by sealing loops and recycling nutrients inside the system. Animal manure, garden trash, and kitchen scraps can all be composted to create nutrient-rich compost that improves soil quality. Greywater systems help with nutrient cycling and water conservation by recycling home wastewater for irrigation. Instead of disposing of garbage in ways that worsen the environment, this idea urges homesteaders to see waste as a resource that may be reintegrated into the system.

The creation of diversified and resilient ecosystems is prioritized in permaculture design concepts. Utilizing and appreciating diversity acknowledges that systems with greater diversity are more resilient to environmental alterations, illnesses, and pests. Homesteaders can attain diversity by cultivating a range of crops, including various plant and tree species, and promoting biodiversity in animal systems. The idea of establishing links between components to improve the general health and production of the system is best illustrated by companion planting, which is the practice of growing mutually beneficial plants beside one another. Permaculture design develops ecosystems that resemble the resilience and stability inherent in natural ecosystems by accepting variation.

The necessity of the transitional zones where several ecosystems converge is highlighted by utilizing margins and appreciating the marginal. Ecotones, as these boundaries are called, are frequently more productive and biologically varied than the individual ecosystems they connect. Homesteaders can benefit by incorporating exciting and useful edges into their designs. For instance, you can plant a range of species and create microclimates that support a diversity of plant and animal life by including hedgerows, swales, or the borders of water features. The homestead's productivity and resilience are increased, and the strategic use of edges enhances biodiversity.

Natural pattern integration is highly valued in permaculture design. This idea effectively answers ecological problems by acknowledging that natural patterns result from millions of years of evolution and adaptation. Homesteaders can construct systems in harmony with nature by observing these patterns and replicating them in their creations. For example, spiral patterns can be used to build garden beds or paths. These patterns can be seen in many natural systems, such as the way leaves are arranged on stems or the form of snail shells. Water distribution system layouts that mimic the branching patterns of trees can maximize water flow. Permaculturists build aesthetically beautiful and functionally practical systems by aligning designs with natural patterns.

The succession concept acknowledges that ecosystems undergo several predictable stages as they change. Homesteaders can emulate natural successional processes in their landscape design by using this idea. For instance, establishing soil fertility and structure through pioneer species and nitrogen-fixing plants paves the way for the emergence of more complex and long-lived species. Resilient and diversified ecosystems might gradually emerge thanks to this deliberate planning. Homesteaders develop landscapes that change with time, becoming more complex and productive as they adjust to shifting conditions by utilizing natural successional processes.

Applying small-scale, intense systems that optimize productivity in constrained locations is also given priority in permaculture design concepts. Homesteaders are encouraged to concentrate on high-yield, low-maintenance techniques that maximize available resources by following the small-scale intensive systems concept. This idea is best illustrated by methods like polyculture, which grows several crops nearby, companion planting, and vertical gardening. By narrowing their focus and using their resources best, homesteaders can produce extremely controllable structures that support sustainability.

Integrating is a fundamental idea in permaculture design, which highlights how system components are interrelated. Homesteaders are encouraged to create systems where each component supports and improves the operations of others by adhering to the integration rather than segregation philosophy. For instance, adding chickens to orchards fosters a symbiotic connection in which the birds enrich the soil with their droppings, forage for pests, and enhance the health of the ecosystem as a whole. Including livestock in rotational grazing systems improves soil fertility through manure deposition and facilitates effective fodder management. Harmonious and mutually supporting ecosystems are produced via permaculture design, which cultivates advantageous interactions between elements.

CHAPTER VII

Harvesting and Preserving

Knowing When to Harvest

On the homestead, harvesting is a significant event that marks the end of months of meticulous preparation, care, and excitement. Securing the benefits of a meticulously maintained garden, orchard, or cattle enterprise necessitates carefully balancing art and science. A thorough awareness of environmental variables, the life cycles of plants and animals, and the subtle differences in the growth of each crop and animal are all necessary for making the right harvesting decisions. In this section, we delve into the science of agricultural knowledge and the artistry of observation to examine the various factors homesteaders must consider when determining the best time to harvest.

The life cycle of the crops or animals being raised

directly affects when to harvest them. It is essential to comprehend the distinct growing stages for plant-based harvests. The best time to harvest fruits and vegetables varies based on the variety and species. Certain crops are optimally harvested when fully developed on the plant, exhibiting vivid hues, a distinct aroma, and effortless detachment from the stem or branch. For some, it could be better to harvest at a little earlier stage—also called the "breaker" or "baby" stage—for culinary purposes, guaranteeing the best flavor, tenderness, and nutritional value.

Regarding grains and cereals, timing is essential to

getting the ideal ratio of ripeness to moisture content. Harvesting too soon can lead to undeveloped seeds and poorer yields; harvesting later carries the danger of higher moisture levels, which can promote mold growth and lower quality. The homesteader can make well-informed decisions about when to bring in the harvest by observing environmental conditions and changes in the crop's color and firmness.

Legumes and pulses also require exact timing to achieve the best flavor and texture. These crops' optimal nutritional content and taste can only be achieved by harvesting them at the proper time when the pods are complete, but they need to be more mature. Homesteaders need to be aware of more extensive environmental cues that indicate when it is best to harvest their crops in addition to the specific traits of each crop. The maturation process is influenced by temperature, humidity, and daylight hours, among other factors. Homesteaders who are skilled at interpreting these indicators can harvest more precisely.

Making judgments about when to harvest adds another complexity related to animal husbandry. The animals' age, weight, and condition are essential for meat production. Young animals usually produce tender meat, although the animals' size and weight at harvest must match the market or the individual's preferences. Homesteaders can manage the time of livestock harvests for best results by knowing the growth rates of particular breeds and modifying feed accordingly.

Egg and dairy production also require precise scheduling. Tracking the patterns of animals that lay eggs helps homesteaders plan for effective egg gathering; knowing the peak lactation periods for dairy animals guarantees maximum milk output. When to harvest these goods is determined by factors including the animals' overall health, dietary requirements, and reproductive cycles, in addition to other factors.

Harvest timing involves more than just biology; homesteaders must also deal with outside influences. The best time to bring in the crop can vary greatly depending on the weather. Rainfall, temperature swings, and prolonged sunshine all affect the growth and quality of crops. Homesteaders in areas with distinct seasons take advantage of this natural cycle to schedule their harvests for maximum benefit. The change from summer to fall signifies the peak ripening phase for many fruits and vegetables. Therefore, homesteaders step up their gathering efforts during this time.

On the other hand, weather patterns can be unpredictable and result in unplanned frosts or extended droughts, which could force changes to harvest schedules. Some crops are vulnerable to frost, so farmers race to harvest before the temperatures drop. On the other hand, prolonged dry spells could necessitate closely monitoring soil moisture levels to avoid early ripening or growth retardation. Every harvest season offers different chances and difficulties in the intricate dance that homesteaders and the natural world perform together.

Harvesting is an artistic endeavor that requires acute perceptive abilities honed via experience and a close bond with the property. The ability to discern subtle signals from their plants and animals, such as color changes, the scent of ripening fruit, and the movements of grazing animals, is essential for homesteaders since it helps them make informed decisions about when to harvest. This observational art takes the whole farmhouse ecology, not just the fields and barns. Various factors, including the diversity of plant species, the quality of the soil, and the presence of helpful insects, influence decisions on when to harvest.

Understanding the unique traits of every crop and animal is a crucial component of this skill. Different varieties may show varied ripening patterns within a single crop, react differently to climatic factors, or need other maintenance procedures. The creativity of homesteaders who save seeds is enhanced when they choose and preserve seeds from plants that exhibit exceptional traits, such as the ideal maturity timing.

Knowing when to harvest in the context of animal husbandry involves comprehending the subtle differences between each species and breed. Since every animal matures differently, an experienced homesteader can provide a more detailed assessment of variables, including muscle growth, fat content, and general health. Completing the circle of animal husbandry for homesteaders is the capacity to determine an animal's suitability for harvest through touch, sight, and even behavior.

Homesteaders can improve their decision-making skills

by using the scientific framework of agriculture, even though the art of harvesting when to harvest is primarily based on observation and intuition. Agronomy, horticulture, animal science, and environmental studies are only a few of the fields that make up scientific knowledge. These fields provide essential insights into the physiological mechanisms controlling the growth of plants and animals.

For instance, in plant science, homesteaders can determine the best time to harvest by comprehending the notion of physiological maturity. When a plant reaches physiological maturity, it has finished developing and is ready to start seedlings. In terms of fruits and vegetables, this stage is when the crop peaks in terms of flavor, texture, and nutrient content. Homesteaders skilled in plant physiology can objectively measure fruit maturity by using instruments like refractometers to assess sugar concentration.

Similarly, the study of animals offers a basis for

deliberating on when to harvest livestock. Homesteaders can forecast the ideal age and weight for producing meat by analyzing growth rates, muscle development, and fat deposition. Nutritional science is essential to create balanced meals supporting targeted growth rates and guaranteeing that animals reach harvest conditions effectively and sustainably.

Canning and Preserving Techniques

Homesteaders may prolong the shelf life of their crops and enjoy the tastes of every season all year round by using time-honored methods like canning and preserving. These techniques, rooted in preserving fruits, vegetables, and other perishables, turn the homestead's fleeting produce into long-lasting jars of jams, pickles, sauces, and more. In this section, we explore the various methods that help homesteaders fill their cupboards with the essence of their harvest as we dig into the art and science of canning and preserving.

Fundamentally, canning is the process of sealing food in jars to keep it from spoiling and becoming contaminated by microorganisms, prolonging the preserved goods' shelf life. This process maintains the nutrients and flavors of fresh vegetables in a dependable and tried-and-true way. Homesteaders have more options for protecting a variety of foods thanks to the two main canning techniques: pressure canning and water bath canning.

A common and easily accessible way to preserve high-acid foods, including fruits, jams, jellies, and pickles, is water bath canning. Jars of prepared food are immersed in a boiling water bath for a predetermined time, efficiently killing bacteria and enzymes that can lead to food spoiling. For many homesteaders, water bath canning is a simple and safe procedure because the acidity of these items makes hazardous germs hostile.

Pressure canning, on the other hand, is necessary to preserve low-acid goods such as soups, meats, poultry, and vegetables. Higher temperatures are needed for low-acid foods to eradicate the possibility of botulism and other dangerous germs. Increasing water's boiling point, a pressure canner enables homesteaders to reach the required temperature for the safe preservation of low-acid foods. This process is critical for low-acidity home-canning products to be safe and high-quality.

Whichever canning method is selected, a few essential processes are usually involved. Homesteaders start by cleaning, peeling, and chopping the food into appropriate portions. The cooked meal is then poured into sterilized jars, leaving a certain amount of room to accommodate expansion during the canning process. Brine, syrup, or sauce additions improve flavor and preserve food better. Homesteaders seal the jars and submerge them in the water bath or pressure canner to allow the heat to reach the inside and expel any air. The sealed jars cool once the canning process is finished, forming a vacuum seal that keeps air and bacteria out.

Preserving techniques include more than just canning; they have a variety of approaches that allow homesteaders to add variety to their pantry offerings. Additional methods for prolonging the shelf life of different foods include freezing, dehydrating, fermenting, and curing.

Fruits, vegetables, and even meats can preserve their freshness and nutritional worth by using the simple technique of freezing. When produce is at its ripest, homesteaders can freeze it to keep nutrients and flavors until needed. Certain fruits and vegetables can be frozen immediately, but others might need blanched beforehand to maintain their texture and color. Foods that have been frozen can retain their quality and avoid freezer burn by using airtight bags or containers.

A tried-and-true method, dehydrating is taking the moisture out of food to prevent the growth of fungus and bacteria that cause spoiling. Homesteaders can dry fruits, vegetables, meats, and herbs using a food dehydrator, an oven, or sun. Not only are dehydrated foods lightweight and convenient to store, but they also hold onto a concentrated flavor that enhances the flavor of rehydrated recipes.

Food transforms during fermentation, improving its nutritional value and preserving it. Fermented foods naturally undergo a preservation process thanks to the activities of helpful microbes like bacteria and yeast. Pickles, kimchi, sauerkraut, and fermented drinks like kombucha are a few examples. Foods that have undergone fermentation have longer shelf lives and acquire distinctive aromas and textures that add to the homestead pantry's culinary diversity.

Curing is a typical technique to remove moisture from foods, especially pork and fish. In traditional methods, a curing mixture is made with sugar, salt, and occasionally other ingredients. The mixture is then applied to the meat, letting it seep in and absorb moisture to prevent the growth of bacteria that cause spoiling. Although curing is sometimes connected to producing goods like bacon or smoked fish, it is a tried-and-true preservation technique that has been used for millennia.

Careful attention to detail, strict adherence to recommended procedures, and knowledge of the particular qualities of each food item are necessary for successful canning and preservation. The choice of fresh, ripe food, the cleanliness and disinfection of equipment, and the observance of suggested processing periods are only a few of the variables that affect the final product's quality. Homesteaders must also consider the storage conditions for preserved commodities, such as airtight containers for dehydrated or frozen foods and cool, dark areas for canned goods.

Beyond the usefulness of food preservation, canning and preserving are appealing because they celebrate seasonal bounty and culinary talent. Through this procedure, homesteaders can freeze-dry their harvest, keeping the food and the tastes and memories of each season. The vivid colors of summer fruits facilitate a sensory tour of the homestead's cycles turned into jewel-toned jams, the savory scent of pickles made with garden herbs, and the comforting images of preserved soups in jars for winter.

In addition to being a delightful culinary activity, canning and preserving supports the more significant ideas of sustainability and self-sufficiency. Using the farm's bounty, people become less dependent on industrially processed, store-bought food, which reduces packaging waste and benefits nearby ecosystems. A helpful tactic for lessening the effects of seasonal fluctuations is to preserve the harvest, which enables homesteaders to continue providing a varied and wholesome meal all year round.

Techniques for canning and preserving food also have cultural value, as they transmit customs and culinary expertise from one generation to the next. Many homesteaders receive family recipes, keeping skills, and tried-and-true procedures honed over many years. Sharing conserved foods with loved ones, neighbors, and friends strengthens the homestead's position as the center of sustainable living by fostering community and connection.

But there are difficulties in the science and art of canning and preservation. Sufficient and effective conservation necessitates close attention to detail, following suggested practices, and continuous education. Food illness that can be fatal, such as botulism, emphasizes the importance of following canning instructions, especially when it comes to low-acid foods. To guarantee the security and caliber of their preserved items, homesteaders need to keep up with the most recent regulations and industry standards.

To sum up, canning and preservation methods are a classic and indispensable part of homesteading. Homesteaders may turn the transient bounty of each season into a lasting legacy of flavors and memories with these techniques, which exemplify the union of art and science. Homesteaders create a larder that showcases the variety and depth of their produce through rigorous methods like water bath canning and creative fermentation. In the jars that border the One jar at a time, the spirit of the farmstead is conserved on the shelves.

Maximizing Yield and Minimizing Waste

The delicate dance between maximizing produce and decreasing waste is at the core of successful homesteading. Finding this balance requires careful planning, resource management, and a thorough understanding of the many linked systems that make up a healthy homestead. This section examines the various tactics and all-encompassing methods used by homesteaders to maximize output and reduce waste, establishing a productive and sustainable ecosystem that is in tune with the natural cycles of the land.

To maximize yield, careful planning and strategic design are essential skills. Before setting out on this adventure, homesteaders carefully examine their land, considering variables like topography, soil quality, water supplies, and microclimates. This fundamental knowledge maximizes the utilization of available space and resources by strategically placing gardens, orchards, and animal enclosures. Homesteaders build effective systems with well-considered designs in which each component enhances and supports the production of others.

Homesteaders have long used companion planting, which increases output by using synergies between several plant species. Homesteaders establish robust and mutually beneficial plant communities by carefully matching crops that promote each other's growth and ward against pests. For instance, growing flavorful herbs like basil beside tomatoes enhances their flavor and is a natural pest repellent. Comparably, intercropping, which involves planting various crops adjacent to one another, maximizes available space and resources while fostering biodiversity and raising total productivity.

Another crucial tactic for increasing output and lowering the danger of insect infestations and soil deterioration is crop rotation. Homesteaders avoid accumulating certain pests and diseases linked to specific plants by dividing their harvests each season among various garden areas or fields. Crop rotation also ensures a sustainable and balanced use of the land by adjusting the nutrient requirements of different crops, which helps preserve soil fertility.

Beyond strategic planting placement, optimizing yield dramatically depends on the health of the soil. To improve soil fertility and structure, homesteaders prioritize implementing regenerative agriculture techniques, including cover crops and organic matter absorption. Legumes and grasses are examples of cover crops that not only prevent soil erosion but also fix nitrogen, improving the general health of the soil. Compost, manure, or other organic amendments increase the soil's capacity to retain water, promote microbial activity, and raise nutrient levels—all critical for optimizing crop yields.

Water management is a crucial component of yield optimization, especially in areas with erratic rainfall patterns. Water-efficient irrigation techniques, such as drip irrigation or soaker hoses, are used by homesteaders to minimize water loss through evaporation by supplying water directly to plant roots. By collecting and storing rainwater for use later during dry spells, rainwater harvesting systems improve water sustainability even more. Homesteaders guarantee steady crop hydration by effectively managing water resources, which promotes vigorous growth and higher harvests.

In animal husbandry, strategic breeding, appropriate diet, and ethical methods are all necessary to maximize productivity. Cattle selectively bred for specific qualities—like high milk yield or high-quality meat—align with the objectives of homesteaders. Rotational grazing methods maximize forage usage by enabling pastures to recover and giving cattle a steady source of nutrient- rich, new grass. Well-balanced meals with homegrown and locally sourced feeds enhance animal productivity and health. This reduces waste and increases the supply of valuable goods like eggs, milk, and meat.

Reducing waste on the homestead requires effective trash management. To establish closed-loop systems that limit the environmental impact of their operations, homesteaders adhere to the reduction, reuse, and recycling principles. For example, composting turns garden waste, kitchen scraps, and animal dung into nutrient-rich compost that closes the nutrient cycle and enriches the soil. Furthermore, mulching techniques are used by homesteaders to prevent weed growth, retain soil moisture, and recycle organic matter back into the environment.

Homesteaders look for creative ways to use every harvest portion to reduce waste and maximize every resource. To maximize nutritional value and minimize kitchen waste, fruit and vegetable peels can be used to make homemade broths. Traditionally regarded as waste, livestock offal can be used to make natural fertilizers or as animal feed. By utilizing resources holistically, waste is reduced, and sustainability is promoted, guaranteeing that every aspect of the homestead contributes to its total output. '

Preservation methods like canning, drying, and fermenting are essential for reducing waste since they increase the shelf life of perishable goods. Homesteaders preserve their excess fruits, vegetables, and herbs for year-round use by turning them into jams, pickles, and fermented treats. Through these preservation techniques, homesteaders can decrease food waste and increase self-sufficiency by building a varied and well-stocked pantry.

Using energy efficiently is another way to reduce waste on the farm. Renewable energy sources, such as solar panels or wind turbines, are operated by homesteaders to power necessary appliances like lighting, warmth, and irrigation. An ecological footprint is lessened by using energy-efficient appliances and methods, such as passive solar building design or energy-efficient lighting. Homesteaders reduce waste related to traditional energy usage and transition to a more resilient and environmentally friendly model by adopting sustainable energy solutions.

Regenerative agriculture emphasizes the improvement and repair of ecosystems using comprehensive strategies that go beyond traditional farming techniques. To build robust and vibrant ecosystems, homesteaders who practice regenerative agriculture prioritize biodiversity, soil health, and sustainable resource usage. Agroforestry is a regenerative strategy that improves productivity while minimizing environmental effects. It involves integrating trees and crops into the same system. Trees create a healthy balance between productivity and sustainability by offering shade, improving soil fertility, and supporting the health of the ecosystem as a whole.

The careful handling of leftovers and byproducts produced on the homestead is included in waste minimization. Using wood chips, straw, and fallen leaves allows homesteaders to create valuable livestock bedding, composting, and mulching resources. Incorporating these leftovers into the homestead's processes keeps everything in check, and the business is more sustainable overall. "

The permaculture philosophy, which emphasizes design principles derived from natural ecosystems, strikes a chord with homesteaders who aim to increase productivity while avoiding waste. Permaculturists study and copy natural patterns to build robust and effective systems that resemble the diversity and abundance observed in natural ecosystems. Permaculture-practicing homesteaders integrate perennial plants, water catchment systems, and beneficial insect habitats to create self-sustaining ecosystems that require little outside input to thrive.

A holistic approach must include education and community involvement to maximize productivity and minimize waste. Homesteaders participate actively in community activities, workshops, and courses to learn from seasoned professionals and share expertise and ideas. The collective knowledge of the homesteading community is enhanced by homesteaders' promotion of a culture of ongoing learning and cooperation, which advances a more significant movement towards sustainable and regenerative practices.

CHAPTER VIII

Embracing a Homesteader's Lifestyle

Connecting with the Seasons

Within the homesteading community, life pulses in rhythm with the varying seasons. The entire fabric of a homesteader's existence is the natural ebb and flow, symbolized by the rhythm of spring blossoms, the richness of summer harvests, the brilliant hues of fall, and the calm reflection of winter. This section delves into the profound significance of connecting with the seasons, examining how homesteaders accept, adjust to, and find inspiration in nature's ever-changing fabric.

The homestead awakens in the spring, a season of hope and rejuvenation. Homesteaders anxiously sow the first seeds in anticipation of the rapidly approaching growing season as the days become longer and the temperature rises. The once-dormant ground bursts to life as buds and blossoms emerge in a brilliant display. The air is filled with the lovely aroma of flowers and fruit trees that burst into a riot of color. Pollinators like bees move from bloom to bloom, coordinating the complex symphony of nature. Homesteaders do the age-old routine of tilling the land to prepare for a bumper crop that will adorn their tables in the upcoming months.

Spring is a time to connect with something more than just planting and nurturing. Homesteaders gain a deep understanding of the fragile natural balance and the interdependence of the plants, soil, and animals that live on their property. When new shoots emerge from the earth in the spring, signaling the promise of future harvests, watching the drama of life develop becomes an observational moment. To make strategic decisions on the upkeep of their land, homesteaders pay attention to the subtleties of weather patterns, temperature swings, and wildlife activity.

The homestead experiences a crescendo of growth and plenty as summer approaches. Gardens are a kaleidoscope of color, with zucchini spreading in a profusion, peppers hanging like ornaments, and tomatoes ripening on the vines. While rows of sunflowers stand tall, soaking up the season's warmth, fruit-laden branches slump beneath the weight of their offerings. On verdant fields, cattle graze, and homesteaders work hard to harvest, preserve, and eat the physical fruits of their labor. There's a constant buzz of activity.

Summer is the best season to experience the fullness of homesteading. A cyclical rhythm of planting, tending, and harvesting envelops homesteaders; it's a dance that reflects the ancient customs of agricultural communities. Long, sunny days make for plenty of outdoor labor possibilities, and hands-on interaction with the soil, plants, and animals strengthens one's bond with the land. Harvesting becomes a festivity, a physical representation of the homestead's life, and the result of months of tending and nurturing.

But there's more to the relationship with summer than just the abundance of food. The warm embrace of the sun, the rich perfume of ripening fruits, and the soft rustle of leaves in the breeze are all sensory pleasures that homesteaders savor throughout the season. Summer turns into a season for neighborhood get-togethers, during which neighbors and friends enjoy the everyday experience of consuming preserves prepared from summer fruits, breaking bread made on the homestead, and celebrating the land's bounty. The relationship with the seasons goes beyond the pragmatics of farming and becomes a celebration of life's cyclical nature and an acknowledgment of the interdependence of all living things.

For homesteaders, transitioning from the warm summer hues to the golden fall palette is a time of introspection and change. Harvest season arrives, offering the fruits of summer labor and a head start on the slower months ahead. Orchards yield an abundance of apples, pears, and persimmons, making them into veritable treasure troves. Squash and pumpkins litter the ground, as the remainder of the veggies are harvested from the garden. The excitement in the fields gives way to the fireside as livestock are ready for the upcoming winter.

Fall captures the spirit of thankfulness on the homestead. Homesteaders express gratitude for the bumper crop, understanding that every vegetable and jar of goodness says their mutually beneficial relationship with the earth. When homesteaders find inventive ways to use every portion of the harvest and waste nothing, the season encourages mindfulness about thrift and resourcefulness. As the leaves fall and the land gets ready for the restorative sleep of winter, autumn also encourages contemplation on the cyclical nature of life. The homesteader recognizes the value of relaxation, introspection, and the cyclical rhythm that permeates all of existence.

Winter comes to the farmhouse with a subdued calm that leaves everything in its wake. The otherwise seemingly inactive natural world is subtly altered when the air crystalizes, and the countryside is blanketed in frost. Homesteaders become more connected to the season as they focus inside, even though the pace of outdoor work may slow down. Winter becomes a time for reflection, organizing, and creative homestead maintenance.

Homesteaders welcome winter as a chance for rejuvenation. Indoor activities take center stage, including planning gardens, browsing seed catalogs, and making handcrafted gifts. As a hub for the room, the fireplace offers warmth and a place for loved ones to congregate. Homesteaders might take a break during the winter to refuel, evaluate the accomplishments and difficulties of the previous year, and plan for the following seasons.

On the homestead, connecting with the seasons is more than just the hands-on aspects of farming and gardening; it's a comprehensive interaction with the natural world. Homesteaders watch the moon and stars dance in the sky, the sun moving through different positions, and the minute variations in daylight. They tune in to the migratory movements of birds, the wildlife's hibernation cycles, and the slow changes in the surrounding flora. This attunement is a visceral, sensory experience that strengthens the bond with the land and creates a strong sense of place rather than just being an intellectual exercise.

Understanding seasonal cycles helps homesteaders decide when to cultivate, harvest, and carry out other essential tasks. Homesteaders acquire the skill of being astute land watchers and an intuitive sense that directs their activities by the natural world's cycles. The ability of the homesteader to time their actions to the cycles of nature is demonstrated by the practice of "winter sowing," in which seeds are placed directly in outdoor containers over the winter to take advantage of the seasonal cues for germination.

On the homestead, regenerative and sustainable methods are also influenced by the relationship with the seasons. Homesteaders try to replicate the cycles of nature in their farming and gardening practices because they understand how nature functions in these cycles. Working with, not against, the natural order is embodied in crop rotation, timed animal grazing, and cover crop use. Homesteaders develop a sense of stewardship by learning about each season's distinctive qualities and ensuring that their operations support the resilience and health of the land. Homesteaders who want to enhance their organic practices find great resonance in the permaculture philosophy, which places a heavy focus on monitoring and responding to natural processes.

Community Building and Sharing Resources

A strong dedication to fostering a sense of community and resource sharing is at the heart of homesteading. Homesteaders embrace a communal attitude that promotes resilience, cooperation, and mutual support, and they understand the interconnectivity of their undertakings, which extends well beyond the bounds of individual plots of land. This section examines the value of community building in homesteading, looking at how homesteaders cooperate to develop strong, enduring bonds by sharing information, resources, and friendship.

Homesteading, sometimes thought of as a lonesome

quest for independence, thrives in an encouraging community. The idea of the lone pioneer is a romanticized ideal, yet cooperation and shared knowledge are essential to a successful homestead. Homesteaders know that creating a solid community requires more than just trading products; it also requires a deep appreciation of the power of working together.

In the context of homesteading, resource sharing is one of the cornerstones of community building. Homesteaders practice reciprocal giving because they understand the tides of abundance that come and go on their property. Fruits, vegetables, eggs, and honey surpluses are distributed to neighbors, forming a network where every participant can access a wide range of fresh, locally produced items. This reduces waste and strengthens the community's overall resilience by guaranteeing that a range of wholesome, locally-grown foods are available to all.

A long-standing custom among farming groups, exchanging seeds demonstrates a dedication to biodiversity and preserving heirloom cultivars. Homesteaders create a rich tapestry of plant diversity by exchanging seeds of flowers, herbs, and vegetables. Sharing seeds is more than just a pragmatic gesture; it symbolizes a heightened awareness of the land's interdependence and the conviction that every seed has a history, a tale passed down through the ages.

Homesteading groups share information and talents in addition to material resources. Community building incorporates workshops, skill-sharing events, and group projects as essential elements. The transfer of expertise creates a communal pool of abilities within the community, whether it is a master gardener conducting a composting workshop, an experienced beekeeper instructing a novice, or a professional carpenter teaching the art of building chicken coops. In addition to empowering individuals, this shared knowledge guarantees the homesteading methods' sustainability for future generations.

Building a community among homesteaders happens in the nearby area and online through social media groups, online forums, and local homesteading associations. The internet sphere offers a forum for exchanging insights, resolving difficulties, and assisting homesteaders who may be distant geographically. An essential complement to the real world, virtual communities help people who share a passion for sustainable living feel community and fellowship.

Another distinguishing feature of homestead community building is shared infrastructure. Homesteaders frequently work together on projects like shared tool sheds, barns, and processing facilities. Community members construct shared spaces that improve efficiency and lessen individual burdens by combining resources, labor, and knowledge. In addition to their practical value, these shared infrastructures symbolize the spirit of community that characterizes homesteading communities.

Collaborations also happen in the field of animal husbandry. To maximize pasture health and avoid overgrazing, homesteaders commonly participate in community grazing arrangements, where livestock is rotated across multiple holdings. In addition to being good for the soil, this approach makes homesteaders feel more interdependent, strengthening the idea that community well-being is increased through collective stewardship.

The resilience of homesteading communities is especially demonstrated during hard times. Homesteaders come together to assist when faced with unforeseen setbacks, insect problems, or extreme weather. Overcoming hardship requires the collaborative use of labor, resources, and expertise. The way the community responds to difficulties reflects a deep-rooted dedication to perseverance and knowledge that the prosperity of one homesteader strengthens the community as a whole.

Homesteading communities are interrelated, as demonstrated by the practice of group gleaning. To ensure the abundance is well-spent, homesteaders gather extra produce from nearby farms or orchards. In addition to addressing the problem of food scarcity, this activity improves communal ties. Gleaning activities frequently turn into happy get-togethers that promote a feeling of unity and a celebration of the richness that nature offers.

The homesteading movement has given rise to the intentional community living philosophy, in which people choose to coexist with shared beliefs and obligations. The values of cooperative decision-making, resource sharing, and sustainable living are embodied in these intentional communities. Members work together on community gardens or renewable energy projects, lending their labor and skills to the cause, forming a small-scale version of the larger homesteading philosophy within a close-knit community.

In homesteading circles, community formation is often aided by education. People can share best practices, discover novel approaches, and learn from one another via workshops, seminars, and casual get-togethers. The community's overall knowledge is enhanced by the dedication to lifelong learning, which promotes experimentation and curiosity. The transmission of customary abilities, like canning, raising animals, or woodworking, guarantees to preserve and continue priceless information within the community.

The homesteading movement emphasizes community development by creating farmers' markets, local food cooperatives, and community-supported agriculture (CSA) initiatives. These online communities offer a way for homesteaders to engage with people outside their immediate homestead by sharing their produce with a broader audience. In turn, this gives consumers access to fresh, locally farmed food and fosters a direct relationship with the farmers, strengthening community and confidence.

Homesteading fosters a sense of community that encompasses social and emotional support systems. Homesteaders gain a lot from the companionship of like-minded people because they frequently deal with the difficulties of solitude and the rigors of rural existence. Events such as potluck dinners, social gatherings, and seasonal festivities are platforms for exchanging anecdotes, joy, and the basic pleasures of a rural lifestyle. A crucial safety net is the emotional support that homesteading communities offer, especially during trying times or difficult personal circumstances.

One concrete example of community development on the farm is the establishment of cooperative purchase agreements. Homesteaders band together to purchase feed, seeds, and other supplies in bulk, using their combined purchasing power to negotiate lower costs. This lowers expenses for each individual while enhancing the community's economic resiliency. The cooperative purchasing strategy demonstrates a dedication to economic and ecological living and a common knowledge that financial resources are best used when combined for the good of all.

In conclusion, the foundation of the homesteading way of life is community development and resource sharing. Homesteaders understand that working with people who share their goals and beliefs enriches and strengthens their efforts. Homesteading communities are linked through tangible exchanges of things and a deeper bond based on mutual support, shared experiences, and a common dedication to sustainability. A robust and colorful tapestry of relationships that endures through the changing of the seasons and the passage of time is cultivated by homesteaders as they fortify and build their communities.

Overcoming Challenges and Celebrating Successes

Homesteading presents a charming image of life next to the soil with its pastoral charm and self-sufficiency promises. But intertwined throughout this picture-perfect way of life are the inescapable difficulties that put homesteaders' grit and will to the test. The path is full of challenges, from erratic weather patterns to pest infestations and a challenging learning curve for those new to the lifestyle. However, the homesteader finds an unbreakable spirit, ingenuity, and a deep connection to the land precisely by conquering these obstacles. This section examines the struggles and victories faced by homesteaders, highlighting the need for resilience in navigating the uncertain route of sustainable living.

The unpredictable nature of the weather is the biggest obstacle facing homesteaders. Climate change has made once-regular seasons even more surprising. Unexpected deluges, extended droughts, or unseasonable frosts can severely disrupt meticulously organized planting plans and test the durability of crops. To cope with such hardships, homesteaders learn to be adaptive. To lessen the effects of unpredictable weather, they diversify their crops, choose hardy types, and practice water conservation. Reading the indications of nature becomes an essential talent for homesteaders, allowing them to manage the constantly shifting dynamics of the environment and make well-informed judgments.

Another major obstacle facing homesteaders pursuing organic and sustainable techniques is pest management. Pests, including insects, rodents, and others, can destroy crops and endanger cattle health. Homesteaders use a combination of natural cures, companion planting, and integrated pest management tactics because they are dedicated to abstaining from synthetic pesticides and herbicides. A thorough grasp of the local ecology is necessary to safeguard crops while maintaining ecological harmony. As a result of their increased observation skills, homesteaders are better able to recognize helpful insects and plants that can support a healthy ecosystem.

The learning curve for people new to homesteading and who have never worked in agriculture can be challenging and even intimidating. It takes commitment and a readiness to take on new challenges to learn skills like sustainable building, soil management, and animal husbandry. Those who homestead frequently find themselves in a never-ending education cycle, actively engaging in the homesteading community, attending workshops, and seeking advice from seasoned mentors. Even if the beginning may be uncertain, the homesteader's dedication to lifelong learning becomes a source of strength and determination.

The physical and mental rigors of homesteading can be overwhelming. Constant physical endurance is required for the daily tasks of tending to gardens, caring for animals, and maintaining infrastructure. The homesteader learns to prioritize tasks and manage their time well, understanding that balance is necessary to avoid burnout. The intrinsic rewards that come from being connected to the land, animals, and the seasonal cycles serve as a source of motivation during trying times.

One particular problem that many homesteaders encounter, especially those who live in rural locations, is isolation. Though peaceful, the dreamlike environment can often make one feel alone. It can be lonely not to have regular social encounters when the closest neighbors might be miles away. In response to this difficulty, homesteaders take an active role in fostering community. There are opportunities for connection, support, and experience sharing through the formation of neighborhood homesteading organizations, attendance at farmers' markets, and participation in internet forums. The journey of a homesteader is not made in isolation but rather in the context of a larger community that recognizes and values the particular difficulties and rewards of this way of life.

For homesteaders, limited resources and infrastructure provide extra challenges. Water conservation techniques and careful planning are necessary to ensure water access, particularly in arid areas. Developing off-grid energy options like wind turbines or solar panels requires technological know-how and an initial investment. Building greenhouses, animal shelters, and other necessary structures involves a combination of practical construction abilities and inventiveness. A multi-talented individual, the homesteader learns to adjust and improvise using available resources to construct a viable and well-functioning homestead.

Particularly in the early stages of setting up a homestead, financial concerns can be a significant source of stress for homesteaders. Financial resources may be strained by the expenses of developing infrastructure, buying property, and acquiring livestock and seeds. In response, homesteaders take up thrifty lifestyle habits, embrace a do-it-yourself mentality, and look into alternate revenue streams. "homesteading on a budget" becomes a tenet, promoting ingenuity, setting priorities, and developing financial resilience.

Homesteaders' challenges take on a time dimension due to the cyclical nature of agriculture. The uncertainty of weather patterns, possible pests, and other external factors that can affect harvests confronts farmers with the anticipated and labor-intensive task involved in planting and maintaining crops. The homesteader feels the vulnerability of depending only on the land for food. However, it is precisely in this fragility that the homesteader can weather setbacks, modify plans, and rejoice in the victories resulting from meticulous land maintenance.

Honoring accomplishments on the farm is a crucial aspect of the voyage and a vital source of inspiration. Every accomplishment becomes a reason to celebrate, be it the first harvest of vegetables grown on-site, the arrival of a fresh litter of chicks, or the completion of a sustainable energy system. The homesteader is proud of the observable results of their labors, understanding that every achievement is a tribute to their commitment and the well-being of the ecosystem they maintain.

In particular, the harvest season is a time for celebration and the conclusion of the homesteader's labors. A profusion of fruits, vegetables, and herbs bloom in fields and gardens, signifying the culmination of months of hard work. Harvesting becomes a ritual, a get-together between neighbors and friends to partake in the produce. When the crop is preserved via canning, drying, or fermenting, it becomes a year-round source of nutrition. Harvest celebrations serve as a collective recognition of the interdependence of the farming lifestyle and a window into individual accomplishments.

The birthing season is another time of hardship and joy on a homestead. Newborn animals, whether lambs, chicks, or kids, signify the cyclical aspect of homestead life. Homesteaders celebrate the continuation of their livestock and the possibilities they offer for future harvests while navigating the difficulties of guaranteeing the health and well-being of newborns. A homesteader's perspective is rooted in a sense of oneness with life's natural cycles.

For homesteaders, infrastructure initiatives like installing a rainwater collecting system or a sustainable energy system provide concrete indicators of success. The accomplishment of these initiatives improves the homestead's sustainability and efficiency and shows how resourceful and dedicated the homesteaders are to caring for their land. Honoring these successes reinforces the sense of empowerment and agency that homesteaders get from actively modifying their living space.

Sharing extra fruit with neighbors and the larger community turns it into a happy way to show off your abundance. Homesteaders contribute to the community via farmers' markets, community-supported agriculture (CSA) initiatives, or face-to-face interactions with neighbors. The neighborhood food system fosters closer ties between people. Sharing encompasses more than just the pragmatics of allocating resources; it also represents the giving and reciprocity that characterize the homesteading community.

CHAPTER IX

Troubleshooting and Problem Solving

Common Issues in Homesteading

The weather can be harsh and unexpected, one of the biggest problems homesteaders face. The romantic view of the seasons developing in perfect harmony frequently clashes with the harsh realities of severe weather, early frosts, or protracted droughts. These variations in the weather have the potential to seriously affect agricultural harvests, jeopardize livestock, and ruin well- laid plans. Adopting an attitude of flexibility, homesteaders intend to lessen the effects of erratic weather. This could be choosing drought-tolerant crop kinds, conserving water, or building greenhouses or other protective structures to safeguard delicate plants.

One of the biggest challenges facing homesteaders who follow organic and sustainable techniques is managing pests. Pests such as rodents and insects can swiftly destroy crops and endanger livestock. Homesteaders choose natural pest control techniques over synthetic pesticides used in traditional agriculture. Establishing a balanced environment may entail companion planting, promoting beneficial predators, and preserving a healthy level of biodiversity on the farm. The ongoing attention to detail needed for pest control strengthens the homesteader's position as a land steward, sensitive to the fragile natural equilibrium in the homestead setting.

Resource limitations are a significant problem for homesteaders, both financially and physically. Financial resources may be strained by the upfront expenditures of developing infrastructure, acquiring land, and buying seeds and livestock. To maintain their way of life, homesteaders frequently adopt a thrifty mentality, embrace a do-it-yourself mentality, and look for alternate sources of income. As homesteaders develop innovative ways to meet infrastructure needs—whether constructing buildings out of salvaged materials or recycling existing resources—resourcefulness emerges as a critical quality.

For people who have never worked in agriculture, the high learning curve of homesteading can be both a hardship and a chance for personal development. Proficiency in soil management, animal husbandry, and sustainable building practices necessitates commitment and an openness to learning from achievements and setbacks. In addition to actively participating in courses, getting advice from seasoned mentors, and attending events, homesteaders also actively participate in the larger homesteading community. The progression from inexperienced to experienced homesteaders is characterized by an ongoing dedication to learning new things and modifying methods.

The homesteading lifestyle has inherent physical and mental challenges. Constant physical stamina is necessary for the daily tasks of tending to plants, caring for animals, and maintaining infrastructure. Homesteaders must learn how to prioritize tasks and manage their time well to balance the demands of daily living. Moments of physical and mental exhaustion might be spurred on by intrinsic rewards derived from labor, such as a connection to the land and tangible results.

One particular problem that many homesteaders encounter, especially those who live in rural locations, is isolation. Though attractive, the peace of the countryside might make one feel alone. It can be lonely not to have regular social encounters when the closest neighbors might be miles away. In response, homesteaders actively look to connect with the community. Engaging in internet forums, farmers' markets and local homesteading groups are essential ways to meet people, get support, and exchange experiences. The journey of the homesteader is not made in a vacuum; instead, it is undertaken in the context of a larger community that recognizes and values the particular difficulties and rewards of the way of life.

For homesteaders, managing and gaining access to water is crucial, particularly in areas with dry weather. Water conservation, effective irrigation techniques, and the development of sustainable water sources become critical. Relying on wells, rainwater collection, or other alternative water sources necessitates careful planning and infrastructure expenditure. The lack of water adds another difficulty to regular homesteading duties, necessitating cautious handling to protect the homestead's sustainability and water resources.

For homesteaders, money worries can be a significant source of stress, especially in the early stages of setting up a homestead. Financial resources may be strained by the expenses of developing infrastructure, buying property, and acquiring livestock and seeds. In response, homesteaders take up thrifty lifestyle habits, embrace a do-it-yourself mentality, and look into alternate revenue streams. "homesteading on a budget" becomes a tenet, promoting ingenuity, setting priorities, and developing financial resilience.

Although it is essential to homesteading, animal husbandry presents a unique set of difficulties. Homesteaders prioritize the health and welfare of their animals, and they deal with challenges, including disease control, sustainable grazing methods, and ethical breeding. In particular, the birthing season calls for extra caution to protect the health of the young animals. Homesteaders frequently use holistic and natural methods to care for their animals. They are lifelong learners who learn new skills to overcome obstacles and give their livestock the best living environment.

For homesteaders, creating and maintaining infrastructure presents constant problems. Building and keeping greenhouses, animal shelters, and other necessary structures requires a creative mind and practical construction knowledge. Many homesteaders take on a do-it-yourself attitude, recycling materials and modifying pre-existing buildings to suit their requirements. Fences, water systems, and other infrastructure need constant maintenance, and homesteaders get skilled at troubleshooting and fixing problems as they come up.

Homesteaders' challenges take on a time dimension due to the cyclical nature of agriculture. The uncertainty of weather patterns, possible pests, and other external factors that can affect harvests confronts farmers with the anticipated and labor-intensive task involved in planting and maintaining crops. The homesteader feels the vulnerability of depending only on the land for food. However, it is precisely in this fragility that the homesteader can weather setbacks, modify plans, and rejoice in the victories resulting from meticulous land maintenance.

In conclusion, homesteading has its share of common problems despite being a very fulfilling lifestyle. Homesteaders confront various difficulties, ranging from financial limitations and environmental influences to the never-ending process of learning and adjusting. Homesteaders, however, find their genuine grit, fortitude, and deep satisfaction in creating a sustainable and self-sufficient life connected to the earth when they embrace these obstacles. The homesteader's narrative is shaped not just by their daily practices but also by their ethos and connection to the ecosystem they steward as a result of their journey through conquering these frequent difficulties.

Natural Pest Control Methods

Homesteaders understand the delicate balance in the ecosystem they maintain, so they are dedicated to organic and sustainable approaches. Managing pests is an ongoing task, and using artificial pesticides goes against the principles of conscientious land management. Homesteaders respond by using environmentally friendly, natural pest control techniques. This section highlights the connections between homesteaders' natural ways and the larger ecology by examining various strategies they use to safeguard their cattle, gardens, and harvests.

Companion planting is one of the most essential homesteading tools for natural pest management. Homesteaders build a diversified and robust ecosystem by mixing crops with plants that repel pests or draw beneficial insects. For example, the smell of marigolds repels nematodes, while nasturtiums attract predatory insects that eat common garden pests. When thoughtfully chosen and positioned, Companion plantings offer insect protection and improve soil fertility and crop health in general.

Promoting biodiversity on the homestead is an effective way to keep pests away. A wide diversity of plants draws different kinds of insects, which balance the environment and allow predators to control pest populations naturally. Aphids and caterpillars are among the frequent garden pests consumed by predatory insects, including ladybugs, lacewings, and predatory beetles. Homesteaders understand that a healthy population of natural predators supports a robust and autonomous environment. Therefore, they actively work to provide habitats for these beneficial insects.

Biological pest control techniques utilize the inherent mechanisms of nature to regulate pest populations. An efficient and greener alternative is to bring native predators to the homestead ecology. For instance, controlling pests like fleas and some kinds of beetles can be achieved by using beneficial nematodes in the soil.

Similarly, releasing predatory insects that lay their eggs on or within nuisance insects—like parasitic wasps— offers a focused and environmentally responsible way to manage pest populations.

Homesteaders use cultural methods that throw off pests' life cycles and lessen their damage to crops. For example, crop rotation is the practice of growing different crops in subsequent seasons to avoid the accumulation of pests that become specialists in particular plants. Because of this disturbance to their life cycles, the pests find it more difficult to build healthy populations. Furthermore, pests need clarification on covering crops and intercropping practices, which provide a dynamic and less conducive environment for pest reproduction.

Homesteaders frequently use neem oil, a natural insecticide made from the seeds of the neem tree. Neem oil is versatile in the homesteader's toolbox because of its insecticidal, fungicidal, and pesticidal qualities. It is the go-to option for people dedicated to maintaining the ecosystem's delicate balance because it interferes with pests' feeding and reproductive cycles while providing minor damage to beneficial insects.

Diatomaceous earth is a physical pesticide safe for animals and people yet lethal to many insects. It is a fine powder derived from fossilized diatoms. Diatomaceous earth works wonders for managing pests like ants, fleas, and beetles by dehydrating and harming the exoskeletons of insects when it is sprayed on plants or soil. Because it is non-toxic, homesteaders looking for chemical-free pest control solutions can't go wrong with it.

By controlling pests and enhancing soil health, composting contributes to natural pest control in two ways. Compost piles with adequate air circulation foster the growth of advantageous microorganisms that inhibit unwanted pests and illnesses. Composting also draws beneficial insects that eat pest larvae and eggs, like ground and rove beetles. Homesteaders who compost their waste improve their soil and create a healthy ecosystem that controls bug populations.

Although handpicking pests is one of the earliest pest control methods available to homesteaders, its efficacy has remained strong. Homesteaders can stop infestations from becoming harmful by routinely checking their plants and physically getting rid of pests. This practical method aligns with the ideas of natural and sustainable homesteading, although it does demand attention.

Homesteaders now have another weapon for natural pest management in the form of essential oils made from aromatic plants. For example, peppermint oil has insect-repelling qualities and can keep pests out of homes and gardens. Similarly, natural insect repellents made with oils like citronella, lavender, and eucalyptus are utilized. Homesteaders utilize their fragrant and deterring properties to incorporate natural solutions such as essential oils into their pest management tactics.

Pests can be kept out of gardens and crops with the help of physical barriers. For instance, row covers are a lightweight cloth that protects plants from pests and flying insects. Similarly, floating row covers let rain and sunlight to the plants while shielding crops from early-season pests. Homesteaders safeguard their crops by erecting these barriers, which lessens the need for more aggressive insect management techniques.

Homesteaders frequently use smell as a deterrent against pests. Because of their inherent scents, fragrant plants like marigolds, basil, and mint can keep pests away. This approach makes the homestead more attractive and diverse and more pest-resistant. The interplay of scents forms an essential component of the homesteader's organic approach to pest management, producing an aromatic and valuable garden.

Integrated Pest Management (IPM) is a comprehensive pest management strategy that combines several natural techniques. IPM strongly emphasizes cultural practices, biological controls, and preventive approaches to manage pests with the least negative environmental impact. Homesteaders build a thorough and long-lasting pest management system that adheres to the values of responsible land care by combining a variety of tactics.

On the homestead, educational programs and community engagement are essential to natural pest management. To promote a shared awareness of sustainable practices, homesteaders actively teach their communities about natural pest management techniques. Workshops, neighborhood gatherings, and internet discussion boards provide venues for sharing knowledge and experience on efficient and green pest management techniques.

Finally, using natural pest management techniques shows how homesteaders are dedicated to balancing their practices with the ecosystems they nurture. A wide range of tactics, such as companion planting, biological controls, cultural customs, and physical barriers, are used by homesteaders with an emphasis on environmental sustainability. The combination of these techniques represents a responsible land stewardship philosophy that aims to establish a robust and well-balanced ecosystem on the homestead, in addition to being a pragmatic approach to pest control. Homesteaders connect with the cycles of nature via their pursuit of natural pest control, understanding that a robust and healthy ecosystem is the cornerstone of an independent and sustainable way of life.

Disease Prevention and Management

When it comes to homesteaders who are interested in developing a lifestyle that is both sustainable and self-sufficient, one of the most critical concerns is ensuring the health and well-being of both their crops and their cattle. To effectively prevent and manage diseases on the homestead, it is necessary to use a holistic approach incorporating preventative measures, vigilant monitoring, and responsive interventions. In this section, we delve into the myriad of tactics homesteaders use to protect their agricultural pursuits. We place particular emphasis on the interdependence of the health of animals and crops within the complicated web that is the homesteading lifestyle.

Maintaining a clean and hygienic environment is one of the key pillars that contribute to preventing sickness on the homestead. Hygiene practices are implemented in animals' habitats and the regions where crops are grown. It is possible to lower the risk of disease transmission and create a healthier living environment for animals by regularly cleaning their barns, coops, and feeding facilities. In the context of crop production, good sanitation entails the removal of debris, the rotation of crops, and the timely removal of unhealthy plants to prevent the spread of diseases. Homesteaders are aware that a clean and well-maintained environment is the foundational component when it comes to preventing the start and spread of infections.

On the homestead, the implementation of strategic animal husbandry practices makes a substantial contribution to disease prevention. It is possible to strengthen cattle's immune systems by providing them with adequate nutrients, access to clean water, and balanced diets. This will make them more resistant to infections. In addition, homesteaders employ quarantine procedures for new animals to prevent the spread of diseases that could affect the herd or flock already in existence. In addition to routine health checks, immunizations, and deworming at the appropriate time, the overall health of livestock is further strengthened. Homesteaders can lessen the likelihood that their animals will contract common diseases by taking preventative measures and adopting a policy of proactive behavior.

Regarding preventing diseases in the agricultural sector, crop rotation and diversity are two of the most critical factors. Monoculture, which refers to the continuous cultivation of a single crop, can produce conditions favorable to the accumulation of particular pathogens and pests. Homesteaders reduce the likelihood of this by rotating their plantings and diversifying their crops. The introduction of cover crops, such as legumes that fix nitrogen in the soil, not only adds to the health of the soil but also disturbs the life cycles of certain illnesses and pests. Crop rotation not only reduces the likelihood of diseases that are transmitted through the soil, but it also helps to create a farming system that is more resilient and sustainable.

Implementing integrated pest management, often known as IPM, is a comprehensive strategy for avoiding diseases in crop agriculture. Homesteaders can construct an environmentally benign and balanced system by utilizing a combination of natural enemies, cultural practices, and preventative measures. Biological pest control is achieved by introducing beneficial insects into the environment, such as ladybugs and predatory beetles. Monitoring crops becomes an essential component of integrated pest management (IPM), which enables homesteaders to detect potential disease outbreaks or pest infestations at an early stage and respond with interventions that are specifically targeted. The homesteader's dedication to environmentally responsible and sustainable land management is aligned with the synergistic integration of numerous integrated pest management (IPM) tactics.

On the homestead, disease control relies heavily on vigilant monitoring and early diagnosis as two of the most important contributing factors. Homesteaders cultivate a sharp eye for tiny indicators of suffering in plants and animals, enabling them to recognize possible health issues before they become more severe. Monitoring the changes in behavior, conducting comprehensive inspections of crops, and performing routine health checks on livestock are all excellent ways to gain valuable insights into the general health of the ecosystem found on a homestead. The homesteader fulfills the function of a caretaker sensitive to the complexities of the natural world, and the focus placed on observation is consistent with this job.

Proactive vaccination tactics are necessary to protect the health of cattle kept on the farm. A collaborative effort between homesteaders and veterinarians develops a vaccination regimen suited to the individual requirements of homesteaders' animals and the diseases that are prevalent in their region. Vaccinations are administered as a preventative strategy to strengthen cattle's immune systems and reduce the likelihood of disease transmission. Homesteaders actively protect their animal populations' long-term health and resilience by actively participating in this collaborative and preventive strategy.

On the homestead, herbal treatments and natural supplements are widely recognized as essential in preventing and managing diseases. Traditional knowledge and herbalism are becoming increasingly popular among homesteaders to address common health problems affecting plants and animals. For instance, the utilization of garlic as a natural antibiotic for cattle or the utilization of herbal infusions for treating particular plant diseases are examples of practices that demonstrate a dedication to holistic and natural methods. The homesteader's tendency toward sustainable and self-sufficient activities is aligned with incorporating these therapies into the overall healthcare approach.

Homesteaders take precautions to prevent the spread of diseases among their animals by prudently using quarantine protocols. To facilitate observation and health evaluations before their incorporation into the current herd or flock, new animals are separated from the rest of the flock for a certain amount of time. By implementing this preventative strategy, the homesteader can reduce the likelihood of introducing infectious diseases and demonstrate their dedication to proactively managing diseases. When combined with regular health examinations, the routine practice of quarantine acts as an essential practice to sustain the overall health and vitality of the animals that reside on the property.

On the homestead, responsible waste management procedures substantially contribute to disease prevention. By way of illustration, the risk of contamination and the spread of pathogens can be reduced by properly disposing of manure that cattle have produced. Homesteaders frequently utilize manure as a vital resource for composting, which results in the creation of a closed-loop system that not only protects against illness but also improves soil fertility. The principles of sustainable agriculture are aligned with the conscientious management of waste, which reinforces the interconnectivity of the health of animals and the health of the land on the farm.

Culturing plant types that are resistant to disease is an example of a proactive approach to avoiding illness in agricultural farming. Homesteaders are noted for their ability to choose and develop plant kinds known for their resistance to diseases in their territory. By decreasing the need for chemical interventions, this technique contributes to creating a more environmentally responsible and sustainable agricultural system. The homesteader's method for constructing a resilient and self-sufficient food production system includes carefully selecting disease-resistant varieties as a crucial strategy component.

Regarding disease prevention techniques for animals on the farm, holistic veterinarian care is a complementary approach. Members of the homesteading community collaborate closely with vets who share their dedication to natural and environmentally responsible procedures. Regular checkups, diagnostic testing, and collaborative talks all lead to the development of an all-encompassing animal medication plan. Not only does holistic veterinary care involve the treatment of specific ailments, but it also emphasizes preventative measures, nutritional guidance, and overall well-being.

In conclusion, the prevention and management of diseases on the homestead require a comprehensive and interrelated approach. It is common knowledge among homesteaders that the health of crops and livestock are inextricably linked. This is the case, whether it is through establishing a clean and hygienic environment, adopting proactive animal husbandry techniques, or utilizing herbal treatments. The homesteader's dedication to responsible land management, environmental harmony, and the long-term resilience of their self-sufficient lifestyle is shown in the adoption of sustainable preventative and proactive actions. As guardians of their ecosystems, homesteaders must traverse the delicate balance between proactive disease prevention and responsive management to cultivate a prosperous and harmonious homestead sustainably

CONCLUSION

Within the lush pages of "Harvest Haven: Creating Your Backyard Homestead Oasis," we have traveled to the core of sustainable living and examined the complex web of abilities, expertise, and enthusiasm that characterizes the homesteading way of life. This e-book, based on converting common areas into flourishing havens of self-sufficiency, is a lighthouse for people who want to develop a closer bond with the land.

"Harvest Haven" develops as a thorough guide, starting with the introductory chapters on understanding and constructing your homestead and continuing with a nuanced analysis of natural pest control, disease prevention, and the rhythms of planting and harvesting. It tells a story of resiliency, flexibility, and the joyful celebration of life near nature, in addition to providing valuable insights into the ins and outs of farming.

The reader is encouraged to reflect on the deep connection between a homesteader and the environment, realizing that every decision made there impacts the surrounding area, the soil, plants, animals, and the community at large. The homesteader appears as a steward of the land, caring for it with wisdom, care, and a strong feeling of duty in the harmonious realm of sustainable living.

"Harvest Haven" presents a comprehensive picture of a way of life beyond basic sustenance as we examine the chapters on crop choices, water systems, tools, and the incorporation of permaculture ideas. It represents a deliberate decision to live, move with the seasons, and create relationships with the community, the environment, animals, and, eventually, oneself.

The theme of empowerment runs through "Harvest Haven" like a narrative thread. It offers individuals the power to regain control over their food, break free from the frantic pace of modern life, and create a haven that is more than just a geographical location- it's also a sanctuary of morals, values, and conscious living. With the aid of this e-book, readers will not only gain valuable knowledge about hoesteading but will also be inspired to embark on a life-altering journey that will enrich their souls as well as their soil.

As we delve into the nuances of building a community, sharing resources, and overcoming hurdles in the final chapters, "Harvest Haven" transcends its role as a guide and becomes a testament to the resilience of the human spirit. In this last act, the homesteader is no longer portrayed as a lone individual battling against the elements, but rather as a member of a larger community—a group of individuals who share common values and unite to exchange knowledge, insights, and the fruits of their labor.

"Harvest Haven" is ultimately a tribute to the potential of creating a bountiful oasis in your backyard, where you actively engage in nature's cycles rather than merely observing them. Here, challenges are met with creativity, and achievements are recognized as significant milestones in a lifelong journey. The reader is left with a wealth of knowledge and an invitation to cultivate the soil, sow the seeds, and watch, with pride and motivation, as the backyard transforms into a thriving homestead oasis as the pages gradually close.

Thank you for buying and reading/listening to our book. If you found this book useful/helpful please take a few minutes and leave a review on the platform where you purchased our book. Your feedback matters greatly to us.

www.ingramcontent.com/pod-product-compliance
Lightning Source LLC
Chambersburg PA
CBHW052051150726

48002CB00002B/840